T0140891

Synthesis and Characterization of Anisotropic Colloidal Particles

DISSERTATION

ZUR ERLANGUNG DES AKADEMISCHEN GRADES EINES
DOKTORS DER NATURWISSENSCHAFTEN (DR. RER. NAT.)
IM FACH CHEMIE DER FAKULTÄT FÜR BIOLOGIE, CHEMIE UND
GEOWISSENSCHAFTEN DER
UNIVERSITÄT BAYREUTH

VORGELEGT VON

Martin Hoffmann

GEBOREN IN MARKTREDWITZ / DEUTSCHLAND

BAYREUTH, 2010

Bibliografische Information der Deutschen Nationalbibliothek

Die Deutsche Nationalbibliothek verzeichnet diese Publikation in der Deutschen Nationalbibliografie; detaillierte bibliografische Daten sind im Internet über http://dnb.d-nb.de abrufbar.

ISBN 978-3-8325-2681-8

Logos Verlag Berlin GmbH
Comeniushof, Gubener Str. 47,
10243 Berlin
Tel.: +49 (0)30 42 85 10 90
Fax: +49 (0)30 42 85 10 92
INTERNET: http://www.logos-verlag.de

Die vorliegende Arbeit wurde in der Zeit von Oktober 2007 bis März 2010 in Bayreuth am Lehrstuhl Physikalische Chemie I unter Betreuung von Herrn Prof. Dr. Matthias Ballauff angefertigt.

Vollständiger Abdruck der von der Fakultät für Biologie, Chemie und Geowissenschaften der Universität Bayreuth genehmigten Dissertation zur Erlangung des akademischen Grades eines Doktors der Naturwissenschaften (Dr. rer. nat.).

Dissertation eingereicht am: 16.03.2010

Zulassung durch die Prüfungskommission: 24.03.2010

Wissenschaftliches Kolloquium: 02.08.2010

Amtierender Dekan:

Prof. Dr. Stephan Clemens

Prüfungsausschuss:

Prof. Dr. Matthias Ballauff (Erstgutachter)

Prof. Dr. Thomas Hellweg (Zweitgutachter)

Prof. Dr. Josef Breu (Vorsitz)

Prof. Dr. Mukundan Thelakkat

ALLES kommt in der Wissenschaft auf ein Gewahrwerden
dessen an, was den Erscheinungen zu Grunde liegt.
Ein solches Gewahrwerden ist bis ins Unendliche fruchtbar.

(Johann Wolfgang von Goethe)

MEINER Familie

Table of Contents

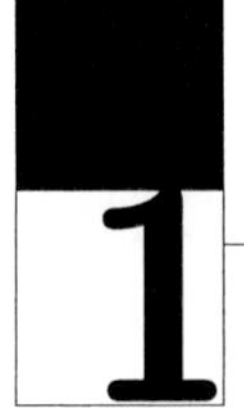

Chapter 1

Introduction

Spherical particles have been thoroughly investigated for the last decades and they are still used as model systems in many theoretical studies. The recent statement a "perfect sphere is no more a winner" clearly expresses the increasing importance of non-spherical colloids [1]. The development of anisotropic colloidal particles with simple [2–5] or complex shape [6–10] and their dynamics in solution are fundamental towards an understanding of many academic and industrial relevant problems like sedimentation, coagulation or rheology [11]. Particles may be anisotropic due to their non-spherical shape, surface chemistry and/or composition (optical anisotropy) [12]. Potential applications of non-spherical particles include for example the tuning of optical properties [13] and the development of optical films which can be used in flat panel displays [14].
Unlike isotropic spheres, the packing configurations of anisotropic particles are influenced by anisotropic directional interactions, as for example shape-selective depletion forces [14, 15]. Theoretical calculations predict numerous novel phases realized through the self-assembly of anisotropic particles [16], like the plastic crystal phase of hard spherocylinders or hard dumbbells [17, 18]. Therefore, for a comparison with model calculations, highly-defined model colloids are needed to correlate the particle morphology (e. g. shape, size) with the microscopic structure of diluted or concentrated suspensions (e.g. crystallinity) and the corresponding macroscopic properties (e.g. flow behaviour) [5, 19, 20].
Depolarized dynamic light scattering (DDLS) is well suited to investigate the dynamics of anisotropic particles in solution as they depolarize light. The dynamics in solution can be quantified in terms of a translational and a rotational diffusion coefficient. Isotropic spheres do not give rise to a DDLS signal. Up to now, DDLS has been applied to colloidal particles ranging from optical anisotropic hard spheres [21] to different non-spherical geometries [22]. However, there is a lack of studies dealing with highly complex structures.
This thesis is centered on novel model systems of anisotropic colloids. The investigation of their dynamics in solution proceeds by means of DDLS for all model systems. The stu-

dies are presented according to the increasing complexity of the colloidal particles. First, results are shown for the hydrodynamic and electrodynamic properties of a spherical polyelectrolyte brush (SPB) with trivalent counterions in an electric field. The results are representative for polyelectrolyte brushes in general. The next contribution explores the motion of dumbbell-shaped polyelectrolyte brushes (DPB) in solution. For this purpose, the bare dumbbell-shaped core particles were investigated by DDLS as a reference system without internal motions prior to the DPB. In addition, it was tested whether SPB give a DDLS signal analogous to the DPB. The third study describes thermoresponsive dumbbell-shaped core-shell particles. Based on these findings, an advanced study with colloidal clusters gave insight how particle morphology is related to translational and rotational diffusion.
A short summary will be given to highlight recent progress in the field of anisotropic particles which is closely connected to the content of this thesis.

1.1 Polyelectrolyte Brushes

Polyelectrolyte brushes result when linear polyelectrolyte chains are affixed to planar [23], cylindrical [24] or spherical systems [25–29]. Polyelectrolyte brushes are subdivided into two classes. If a strong polyelectrolyte, like poly(styrene sulfonate), is grafted to a surface, a *quenched brush* results and the dissociation of ion pairs along the chains is not influenced by the pH-value of the solution [23, 30]. In the case of a weak polyelectrolyte, like poly(acrylic acid), the degree of charging of the monomeric units depends on the local pH-value (*annealed brush*) [23, 30]. The term brush implies that the linear dimensions of the attached chains are considerably larger than the distance between two neighbouring chains on the surface [30]. The main feature of polyelectrolyte brushes is the confinement of the major fraction of counterions in the brush layer which neutralize the electrical charges of the polyelectrolyte chains within the brush [31]. In water, the resulting osmotic pressure governs the stretching of the polyelectrolyt chains (*osmotic brush*). Adding monovalent salt reduces the strong electrostatic interaction within the brush layer (*screening*) accompanied with a deswelling of the brush (*salted brush*). The degree of swelling or deswelling results from the balance of an electrostatic repulsive force which will stretch the chains and a retracting force resulting from the elasticity of the polymer chains. One main focus of this thesis is to extend previous work on spherical polyelectrolyte brushes (SPB) [25, 27, 32, 33] to dumbbell-shaped particles. The following paragraphs introduce different preparation pathways for the SPB as well as their solution behaviour. SPB result when polyelectrolyte chains are densily affixed to a spherical (polymer) core. Figure 1.1.1a shows a sketch of a SPB which is characterized by the core radius R_c, the thickness of the brush layer L, the contour length of the grafted chains L_c and the number of chains per unit area σ (grafting density). SPB can be synthesized *via* photoemulsion polymerization (see Figure 1.1.1a). This grafting-from technique facilitates a dense polyelectrolyte layer chemically bound to the substrate. Well-defined poly(styrene) cores were obtained by conventional emulsion polymerization and covered with a thin layer of a photoinitiator. Upon irridiation with UV light, the initiator decomposes into two radicals which initiate the growth of polyelectrolyte chains on the particle surface and in solution (see Figure 1.1.1b) [25, 26].

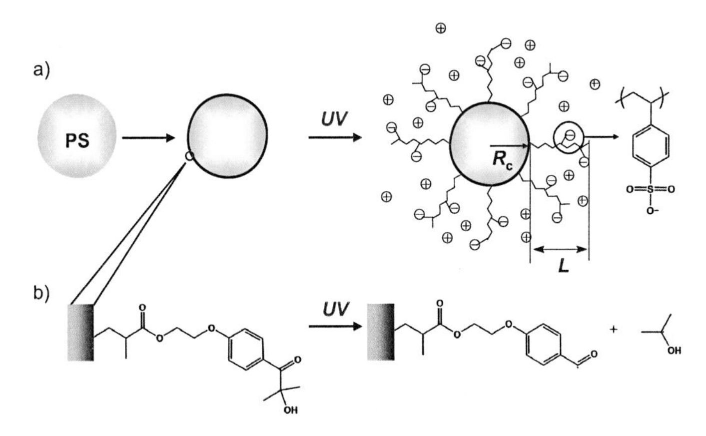

Figure 1.1.1: a) Schematic representation of the synthesis of SPB particles. First, a poly(styrene) core is synthesized in a conventional free radical emulsion polymerization and covered with a thin layer of photoinitiator (HMEM). Upon irradiation with UV light, long chains of a water soluble monomer (e.g. sodium styrene sulfonate) are grafted from the surface of the core particles. b) Decomposition of the photoinitiator HMEM into two radicals upon irradiation with UV light.

The obvious drawback of photoemulsion polymerization is the polydispersity of the grafted chains as $\overline{P_w}/\overline{P_n} \approx 2$. Alternative routes make use of macromonomers [34], controlled radical polymerization techniques [35, 36] or the adsorption of block copolymers with hydrophilic charged and hydrophobic uncharged blocks onto latex particles [37]. The solution behaviour of SPB in the presence of monovalent counterions is governed by the high osmotic pressure within the brush layer resulting from the confined counterions in the brush [30, 38]. However, upon the addition of di- [27] or trivalent counterions [39, 40] the collapse transition of the brush layer is much more pronounced and takes place at considerably lower ionic strength. This can be understood, as the exchange of mono- by trivalent counterions in the brush is entropically favoured. Concomitantly, due to the drastic reduction of the osmotic pressure the SPB flocculate. These experimental findings (see Figure 1.1.2) were corroborated by a simple meanfield approach in an almost quantitative manner by considering the free energy of an isolated SPB as the sum of electrostatic, polymer and entropic contributions of the counter- and coions [38].

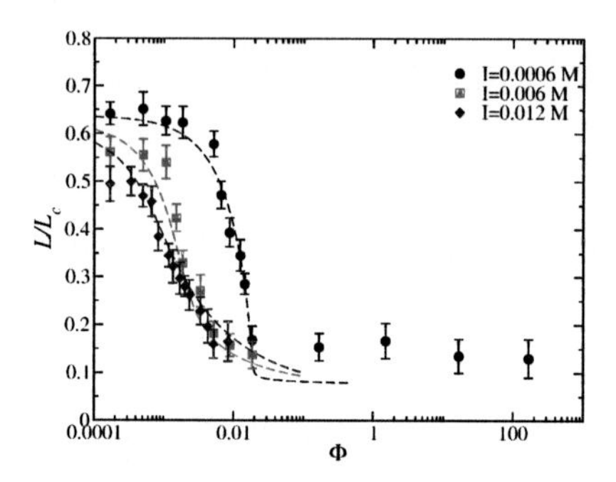

Figure 1.1.2: Brush thickness L normalized by the contour length L_c of a spherical polyelectrolyte brush in the presence of a mixture of mono- and trivalent counterions ($\Phi = [La]^{3+}/[Na]^{+}$) for different ionic strength I. Symbols denote experimental data, and the dashed lines are the result of variational free energy calculations. Reprinted figure with permission from [39]. Copyright 2006 by the American Physical Society.

The dense polyelectrolyte layer leads to both an electrostatic and steric stabilization of the brush. This is a prerequisite to use them as carriers for catalytically active nanoparticles [41–43] or to immobilize enzymes [44]. Recently, the coagulation kinetics of SPB in the collapsed state at higher concentrations of trivalent salt was investigated [45]. The experimental effective surface potential was in very good agreement with theoretical values calculated from the effective particle charge Q^*. Q^* was determined by using a variational free energy calculation [38]. The findings indicate a rather low number of free counterions (*osmotic coefficient*) in the range of 0.05% which is at least one order of magnitude lower as expected from osmotic measurements [46].
Compared to dynamic light scattering, which allows the precise measurement of the brush thickness, anomalous small-angle X-ray scattering (ASAXS) reveals the distribution of counterions inside the brush layer. Dingenouts *et al.* showed that almost all monovalent counterions are trapped in the brush layer. The Manning fraction of the counterions was closely correlated to the polyelectrolyte chains [47]. This experimental findings are in full accordance with theoretical predictions. Interestingly, in ASAXS and SAXS measurements, the SPB exhibited a weak scattering contribution due to the fluctuations of the polymer chains on the surface [47, 48].
The aforementioned studies deal with SPB in the undisturbed state, meaning in the absence of an electric field. However, an electric field may lead to electroosmotic flows inside the brush layer and thus influences the dynamics of the counterions [49, 50]. Zimmermann *et al.* derived from streaming potential measurements that the mobility of monovalent counterions in a planar poly(acrylic acid) brush was markedly lower compared to free ions in solution but lead to enormous surface conductivities in the brush [51].

1.2 Anisotropic Particles with Simple Shape

1.2.1 Colloidal Dimers

Several strategies are available for the fabrication of dumbbell-like particles with a size between 100 nm and several microns (see Figure 1.2.1). For the seeded emulsion polymerization technique developed by Sheu *et al.* [52] crosslinked seed particles are swollen with a hydrophobic monomer at low temperatures prior to the free radical polymerization at elevated temperatures. With this technique, well-defined particles can be prepared with high yields which is suitable for the investigations of bulk properties. The phase separation between the seed particle and the expelled monomer due to the shrinking of the crosslinked polymer network leads to the formation of binuclear colloids (see Figure 1.2.1a).
The Gibbs free energy difference of a monomer within a crosslinked polymer network and a monomer droplet, $\Delta\overline{G}_{m,p}$, can be written as the sum of three contributions: the mixing of monomer and polymer $\Delta\overline{G}_m$, an elastic contribution of the polymer network $\Delta\overline{G}_{el}$ and the interfacial tension between water and the particle $\Delta\overline{G}_t$ [52, 53]:

$$\Delta\overline{G}_{m,p} = \underbrace{RT[\ln(1-v_p)+v_p+\chi_{mp}v_p^2]}_{\Delta\overline{G}_m} + \underbrace{RTN_cV_m(v_p^{1/3}-1/2v_p)}_{\Delta\overline{G}_{el}} + \underbrace{2V_m\gamma/a_{CL}}_{\Delta\overline{G}_t} \quad (1.2.1)$$

R is the gas constant, T the absolute temperature, v_p the volume fraction of polymer in the swollen particle, χ_{mp} the monomer-polymer interaction parameter, N_c the effective number of chains in the network per unit volume, V_m the molar volume of the monomer, γ the interfacial tension between water and the particle and a_{CL} the radius of the swollen particle. As $\Delta\overline{G}_{el}$ and $\Delta\overline{G}_t$ make positive contributions to $\Delta\overline{G}_{m,p}$, the seed particle contraction and thus the formation of the dumbbell morphology, is favoured, while $\Delta\overline{G}_m$ makes a negative contribution to $\Delta\overline{G}_{m,p}$ and promotes seed particle expansion. For micron-sized particles, the interfacial energy term can be neglected compared with the other terms [52, 53]. However, micron-sized particles undergo sedimentation. Therefore strategies are needed to combine both shape anisotropy and colloidal stability.
As one way to address this issue, Mock *et al.* obtained anisotropic nanoparticles with ca. 500 nm size by controlling the surface affinity of crosslinked poly(styrene) seed particles to styrene monomer by coating the surface with a hydrophilic polymer layer [53]. It was found that the amount of surface coating and the crosslinking density determine the degree of phase separation. Moreover, the seeded emulsion technique is capable to fabricate amphiphilic dissymmetric colloids [12]. Kim *et al.* prepared poly(styrene)/poly(methyl methacrylate) (PS/PMMA) or PS/poly(n-butyl acrylate) particles. Such "colloidal surfactants" assemble at liquid-liquid interfaces and stabilize emulsions [54].
Another approach for the two-step synthesis of hybrid dissymetrical nanoparticles was published by Reculusa *et al.* (see Figure 1.2.1d). Small silica seed

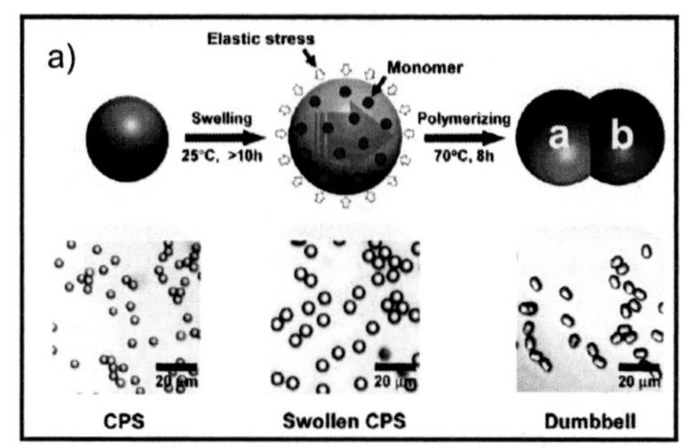

Kim *et al.* (*J. Am. Chem. Soc.* **2006**)

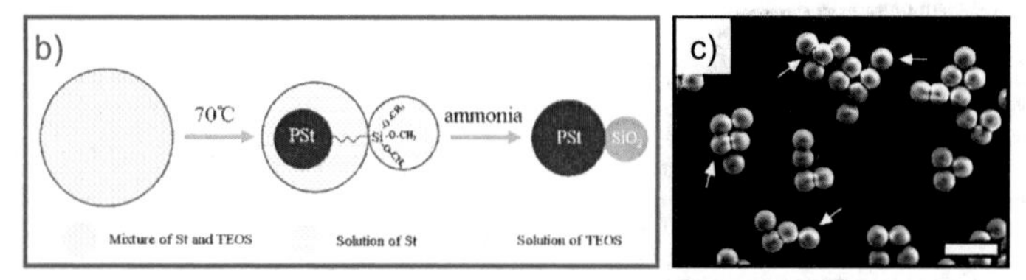

Lu *et al.* (*J. Colloid Interf. Sci.* **2008**) Johnson *et al.* (*Langmuir* **2005**)

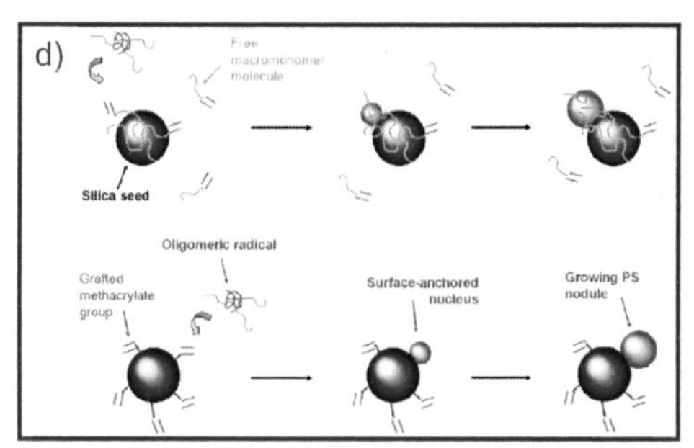

Reculusa *et al.* (*Chem. Mater.* **2005**)

Figure 1.2.1: Overview of different techniques to synthesize dumbbell-shaped colloidal particles. a) Seeded emulsion polymerization with monomer swollen crosslinked seed particles. Reprinted with permission from [12]. Copyright 2006 American Chemical Society. b) One-step miniemulsion polymerization with phase separation of the hydrophobic poly(styrene) core and the hydrophilic silica bulb. Reprinted from [55], Copyright 2008, with permission from Elsevier. c) Controlled aggregation of silica spheres leads to dumbbells. The arrows denote single, triple and quadrupole particles before purification. Reprinted with permission from [2]. Copyright 2005 American Chemical Society. d) Two-step seeded emulsion polymerization of a surface modified silica core with a hydrophobic monomer. Reprinted with permission from [3]. Copyright 2005 American Chemical Society.

particles (50 – 150 nm) were modified by covalently grafting reactive methacrylate groups prior to the free radical emulsion polymerization with styrene [3].
Similar organic-inorganic hybrid dimers were prepared in a one-step miniemulsion polymerization recently [55]. As is schematically shown in Figure 1.2.1b, the reagents were confined in miniemulsion droplets. After copolymerization of styrene with a siliane coupling agent (MPS), the addition of ammonia leads to hydrolysis and polycondensation of tetraethoxysilane with MPS, which results in a phase separation of PS and silica. With this technique, the yield of particle dimers is more than 40%.
In addition of the three (seeded) emulsion polymerizations described above, well-defined colloidal dumbbells can be fabricated by controlled aggregation of two spherical units. For example, Johnson *et al.* described the destabilization of a silica particle suspension and the subsequent shell coating to obtain dumbbells of different aspect ratios [2] . The authors discuss two mechanisms of destabilization, the shear induced coagulation of the silica spheres at high ionic strength and a depletion effect which may be caused by surfactant micelles (Figure 1.2.1c). The fraction of dumbbells was increased from 20% to almost 100% by density gradient centrifugation or repeated centrifugation.

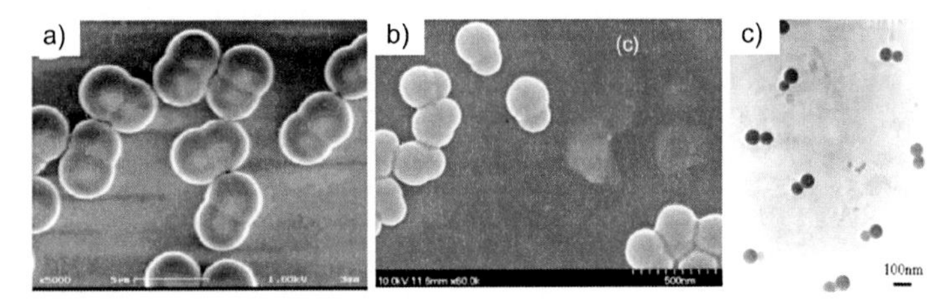

Figure 1.2.2: Scanning electron microscopy image of colloidal dimer particles obtained a) without surface modification of the polymer seed particles. Reprinted with permission from [12]. Copyright 2006 American Chemical Society. b) After coating the polymer seed particle with a hydrophilic polymer layer. Reprinted with permission from [53]. Copyright 2006 American Chemical Society. The scale bars in a) and b) are 5 μm and 500 nm. c) Transmission electron microscopy image of hybrid dissymmetric colloidal particles made of silica (dark grey) and poly(styrene) (grey). Reprinted from [55], Copyright 2008, with permission from Elsevier.

Unlike spherical particles, which are often limited to form crystals with body centered or face centered cubic structures, shape-anisotropic particles can have different microstructures, especially for volume fractions ϕ at the disorder/order phase boundary [20]. Using ultra-small angle X-ray scattering and scanning electron microscopy, Mock *et al.* identified a rotator or plastic crystal phase for $\phi < 0.45$ with a positional order of the centers of mass, but no directional order. This was observed for even higher volume fractions which lead to the conclusion of a body-centered tetragonal phase. An example for the 2-dimensional self-assembly of colloidal dimers has been described by Hosein *et al.* [56]. The authors reported both the experimental finding and the Monte Carlo simulation of a rotator phase for pear-shaped particles with a short bond length.

The examples show a variety of preparation pathways for solid colloidal dimers. However, it would be interesting to attach a stimuli-responsive shell to the dumbbell-shaped core particles in order to adjust the particle properties by changing for example the ionic strength or the temperature.

1.2.2 Ellipsoidal Particles

Colloidal particles with ellipsoidal shape serve as important model systems in condensed matter physics and materials chemistry due to their packing behaviour. Well-defined particles can be obtained at small quantities by embedding spherical particles in a polymer matrix and subsequent stretching of the polymer film [57, 58]. For the fabrication of larger quantities, Sacanna *et al.* stabilized spindle-shaped hematite cores (α-Fe_2O_3) with poly(vinylpyrrolidone) and subsequently covered the core particles with a silica shell in a seeded sol-gel polymerization of tetraethoxysilane (TEOS) (see Figure 1.2.3d).

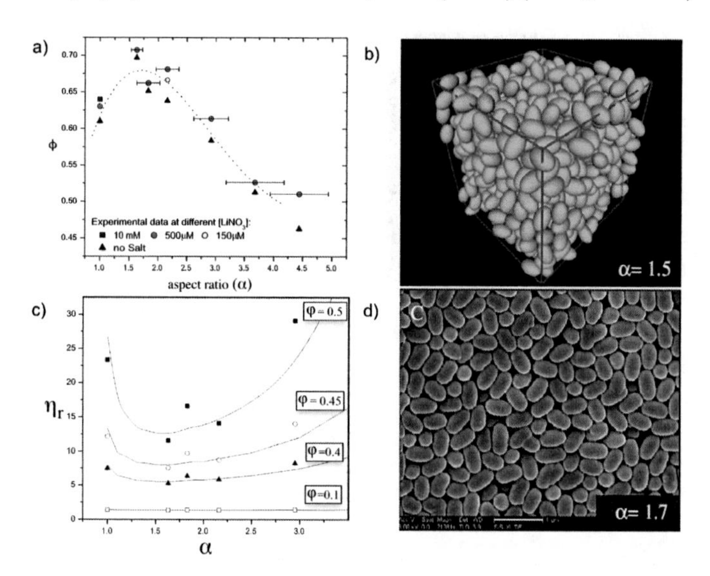

Figure 1.2.3: a) Particle volume fraction ϕ as a function of the aspect ratio α for randomly packed silica ellipsoids at different ionic strength. b) Snapshot of particles simulated by mechanical contraction for $\alpha = 1.5$. c) Relative viscosity calculated for the maximum packing densities from Figure 1.2.3a for different values of α. The symbols denote experimental values, the lines are the result from computer simulations. d) SEM image of a real packing of silica ellipsoids with $\alpha = 1.7$. Reprinted with permission from [5]. Copyright 2007 Institute of Physics.

This approach facilitates different aspect ratios α (length of the long to the short semiaxis) since the silica layer thickness can be adjusted [59]. After rapidly quenching dispersions of ellipsoidal particles with different aspect ratio α by centrifugation, the authors found a maximum of the particle volume fraction ϕ around $\alpha = 1.6$ (Figure 1.2.3a). This experimental findings were simulated by mechanical contraction, a simulation technique for creating dense random packings of hard particles [18] (see Figure 1.2.3b). Interestingly, the density maximum in Figure 1.2.3a is connected with a pronounced minimum of the relative visosity η_r at high volume fractions as shown in Figure 1.2.3c. As the results suggest for ellipsoidal colloids with an aspect ratio α near one in a glassy phase, a small change of α (at the same volume fraction) may cause the melting of the glass due to the reduced viscosity.

1.2.3 Microgels

Microgels are spherical particles made of covalently crosslinked hydrophilic polymer with a size between 50 nm and 5 μm [60]. The most studied example, poly(N-isopropylacrylamide) (PNIPA) undergoes a volume phase transition in water above its lower critical solution temperature (LCST) at approximately 32 °C (thermoresponsive microgel). Concomitantly, the hydrogen bonds between the amide groups and water are weakened and free water molecules are expelled from the shrinking polymer phase [61]. Below the LCST in the swollen state, the microgel is colloidally stable, even at high ionic strength [62]. Figure 1.2.4 shows the volume phase transition of a thermoresponsive microgel network in a schematic fashion.
Microgels may be applied as biosensors [63], delivery systems for biomolecules [64] or photonic devices [65]. However, combining the colloidal stability of a microgel network with a core-shell approach will improve the stablity of the core-particles and lead to materials with novel properties. Spherical core-shell microgels have been fabricated by conventional emulsion- [66] or photoemulsion polymerization [67] using poly(styrene) cores as seed particles (see Figure 1.2.4). Styrene was copolymerized with small amounts of NIPA in the first step to improve the stability of the core particles and to facilitate the growth of the PNIPA network onto the cores in the seeded growth step. As the crosslinking agent N,N'-methylenebisacrylamide (BIS) is polymerized more quickly than NIPA monomer, it is very likely that the crosslinking density is not homogeneous in the network [62, 68]. In general, the core-shell concept is not restricted to the spherical morphology, but no attempts have been made in order to obtain different geometries. Hellweg *et al.* used an elegant one-step synthesis to obtain core-shell microgels with tunable hardness by copolymerization of NIPA with varying amounts of styrene [69]. The first core-shell microgel with a core made of surface modified silica particles and a crosslinked PNIPA shell has been reported by Karg *et al.* [70].
Core-shell microgels were used as stable carriers for noble-metal nanoparticles with potential applications as catalysts [71] or optical materials with a refractive index peri-

odicity on different length scales [72]. Beside these applications, microgels serve as ideal model systems in fundamental research. Crassous *et al.* studied the flow behaviour at the liquid-solid transition of concentrated colloidal dispersions [73]. Hellweg *et al.* investigated the internal dynamics in crosslinked thermosensitive microgels by using neutron spin-echo spectroscopy [74]. It was found that below the LCST the collective diffusion coefficient assigned to a "network breathing" linearly decreased with the crosslinking density. However, above the LCST, the dynamics in the network are frozen and the microgels behave as solid spheres rather than soft colloids [75, 76]. Very recently, by using depolarized dynamic light scattering, Potenza *et. al.* found a small optical anisotropy of spherical core-shell microgels [77]. A reasonable explanation is the inhomogenity of the crosslinked polymer network.

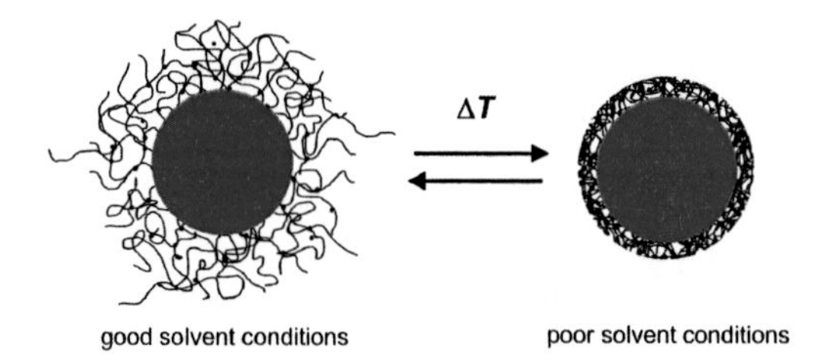

Figure 1.2.4: Sketch of the volume phase transition of a thermoresponsive core-shell microgel.

1.3 Anisotropic Particles with Complex Shape

Colloidal clusters may serve as building blocks for novel colloidal assemblies due to their 3-dimensional complexity. Pine *et al.* described a versatile approach to prepare highly-defined micron-sized clusters made of monodisperse crosslinked poly(styrene) spheres. The underlying self-organization of the particles confined at the oil-water interface in emulsion droplets is depicted in Figure 1.3.1a. Depending on the size of the droplets, a certain number of spherical building blocks is arranged. Evaporation of the oil phase forces the packing of the particles through capillary forces. The final configuration is determined by attractive van der Waals forces. For a small number N of building blocks ($N < 12$), the shape of the clusters corresponds to packings with a minimum of the second moment of the mass distribution (see Figure 1.3.1b)

$$\mathcal{M}_2 = \sum_{i=1}^{N} |\vec{r}_i - \vec{r}_0|^2 \tag{1.3.1}$$

where $\vec{r}_i$ is the position of the center of a building block and $\vec{r}_0$ is the center of mass of the cluster [6, 14].

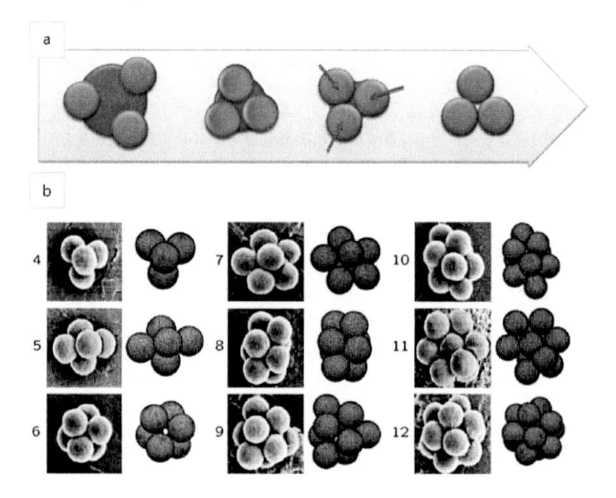

Figure 1.3.1: a) Sketch of self-organizing particles bound to an emulsion interface during the formation of colloidal clusters [14]. Reproduced by permission of The Royal Society of Chemistry. b) Comparison of the experimentally observed packing configurations (left) with results from numerical simulations for $N = 4 - 12$. Reprinted with permission from [78]. Copyright by the American Physical Society.

According to van Blaaderen, colloidal clusters can be termed "colloidal molecules" if their shapes resemble space filling models of molecules [79]. To investigate the complex behaviour analogous to low molecular weight compounds and beyond, it is important to use clusters composed of different materials. Therefore, the approach of Pine *et al.* was extended to building blocks comprised of PMMA or silica [80, 81]. A large number of structural motifs was realized by using binary mixtures of micron- and nanosized particles [8]. To overcome the problem of low yields of monodisperse clusters due to the polydisperse emulsion droplets, Zerrouki *et al.* slightly sheared the emulsion during the cluster formation in a Couette apparatus [82].
Well-defined planar colloidal clusters in the micron-range were used as model systems to study the connection between translational and rotational diffusion at an interface. It was found that the relations could not be described using the Stokes-Einstein-Debye equations for spheres [83]. The data for linear clusters with $N = 2$ to $N = 4$ spherical building blocks strongly suggest a coupling of translational and rotational motion of long linear particles [84, 85]. Up to now, the dynamics of stable colloidal clusters in solution have been unexplored. However, this would be beneficial to investigate the true 3-dimensional diffusion without wall-effects.
The main drawback of micron-sized clusters is the instability of the suspensions due to

sedimentation. Wagner *et. al.* presented an elegant solution of this problem by combining basic principles of miniemulsion polymerization and the aforementioned emulsion encapsulation. Ultrasonication emulsification lead to a reduced size of the emulsion droplets and thus smaller clusters in the range of 300 nm [9].
As another approach to fabricate binary colloidal clusters, Perro *et al.* further developed their previous work on hybrid dissymmetric particles [3, 86]. By adjusting the ratio of monodisperse silica seed particles and attached poly(styrene) "nodules", regular bi-, tri-, and tetrapods were obtained in good yields. As these regular structures can be compared to the space filling models of $BeCl_2$, BF_3 and CH_4 molecules, they are termed *colloidal molecules*. Recently, Kraft *et al.* presented a straighforward one-step synthesis to obtain asymmetric, uniform colloidal molecules which are similar to colloidal water or ammonia [10, 87]. In principle, merging of liquid protrusions attached to crosslinked poly(styrene) spheres leads to the novel structures as shown in Figure 1.3.2.

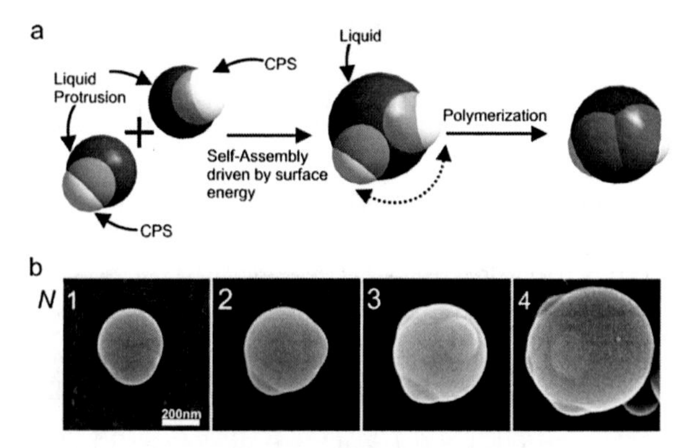

Figure 1.3.2: a) Crosslinked poly(styrene) (CPS) colloids are swollen with monomer to give liquid protrusions upon heating. The fusion of the protrusions is driven by minimization of the surface energy between the monomer and the water phase. b) Scanning electron microscopy images of the particles after polymerization. Assemblies with two or three seed particles resemble water and ammonia-like molecules [87]. Reproduced by permission of The Royal Society of Chemistry.

Many of the examples may be modified by attaching a shell layer to the core structure. This is useful to improve particle stability against coagulation or sedimentation and to fabricate novel materials with different properties. To fulfill these requirements, polyelectrolyte chains or a crosslinked thermoresonsive polymer network were attached to dumbbell-shaped core particles within the framework of this thesis.

1.4 Depolarized Dynamic Light Scattering

The colloidal particle systems presented in this thesis are optical and/or shape anisotropic and thus they depolarize light. Therefore, depolarized dynamic light scattering (DDLS) is well suited to investigate not only the translational but at the same time the rotational diffusion, which is hard to measure in a conventional DLS experiment. Figure 1.4.1 shows the light scattering configurations for a DLS (vV) and a DDLS experiment (vH) schematically.

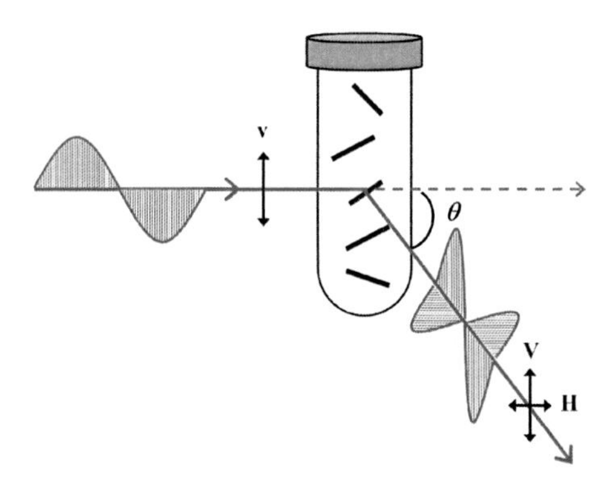

Figure 1.4.1: Schematic representation of the light scattering configurations for a polarized (vV, DLS) and a depolarized dynamic light scattering experiment (vH, DDLS). When particles (size exaggerated) are optical or shape anisotropic they depolarize light.

The general treatment of dynamic light scattering from large anisotropic particles is quite complex, especially in the case of internal degrees of freedom [21, 85, 88]. The following paragraphs review two important cases for a basic understanding.
The light wave in rigid non-spherical particles composed of optical isotropic segments observes a change in polarizability due to translational motion of the particle as a whole, and also due to the relative motion of the centers of the segments. Pecora gave the following expression for the dynamic form factor $S(q,t)$ of a thin rigid rod which is proportional to the normalized field-field autocorrelation function $g^{(1)}(q,t)$ [22, 85, 89, 90]:

$$g^{(1)}(q,t) \propto S(q,t) = S_0(qL)\exp\left(-D^T q^2 t\right) + S_1(qL)\exp\left[(-D^T q^2 + 6D^R)t\right] + ... \quad (1.4.1)$$

where t is the time, q is the absolute value of the scattering vector, L is the length of the rod, $S_0(qL)$ and $S_1(qL)$ are the scattering amplitudes as shown in Figure 1.4.2, and D^T and D^R are the translational and the rotational diffusion coefficient, respectively.

Figure 1.4.2 examplifies that the exponential term with amplitude S_1 contributes less than 1% to the total scattering intensity for $qL < 3$ for purely shape anisotropic particles. Therefore analyzers with a high extinction ratio are needed to obtain the weaker DDLS signal with amplitude S_1 exclusively.

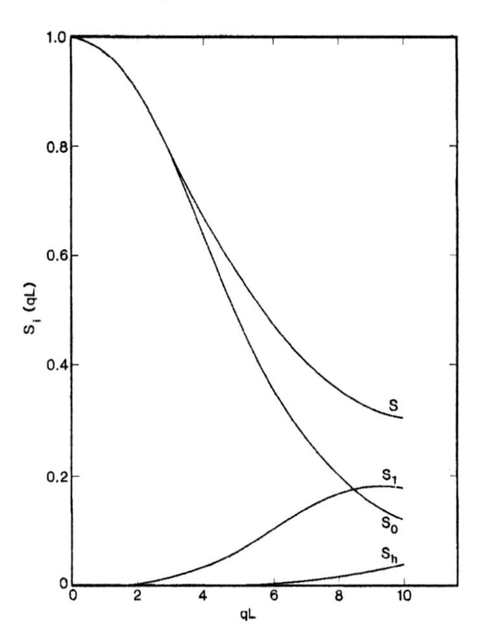

Figure 1.4.2: Scattering amplitudes $S_i(qL)$ for a thin rigid rod. S is the total integrated intensity form factor, S_0 is the scattering amplitude of the pure translational part and S_1 is the amplitude of the first part depending on both the translational and the rotational diffusion. q is the absolute value of the scattering vector, and L the length of the rod. Note that $S = S_0 + S_1 + S_h$ [89]. For comparison, qL is 1.4 or 3.7 for $L = 200$ nm for a scattering angle θ of 30 ° or 90 ° ($\lambda = 632.8$ nm and $n = 1.333$).

If optical anisotropy comes into play, the following expressions hold for monodisperse, cylindrical symmetric particles in highly diluted solutions. The electric field autocorrelation functions $g_{vV}^{(1)}(q,t)$ (DLS) and $g_{vH}^{(1)}(q,t)$ (DDLS) read [85]

$$g_{vV}^{(1)}(q,t) \propto < N > \alpha_{iso}^2 \exp\left(-D^T q^2 t\right) + \frac{4}{45} < N > \alpha_{aniso}^2 \exp[-(D^T q^2 + 6D^R)t] \quad (1.4.2)$$

$$g_{vH}^{(1)}(q,t) \propto \frac{1}{15} < N > \alpha_{aniso}^2 \exp\left[-(D^T q^2 + 6D^R)t\right] \quad (1.4.3)$$

where $< N >$ is the average number of particles in the scattering volume, $\alpha_{iso} \equiv 1/3(\alpha_{\parallel} + 2\alpha_{\perp})$ is the isotropic part of the polarizability tensor and $\alpha_{aniso} \equiv (\alpha_{\parallel} - \alpha_{\perp})$ is the optical anisotropy. $\alpha_{\perp}$ and $\alpha_{\parallel}$ are the polarizabilities perpendicular and along the particle main axis. Examples for optical anisotropic objects are partial crystalline polymer spheres [21] or hybrid particles. The relaxation rate for the DDLS experiment $\Gamma_{vH} = \Gamma_{\text{fast}}$ follows from the last term in eq. 1.4.1 and eq. 1.4.3

$$\Gamma_{vH} = \Gamma_{\text{fast}} = D^T q^2 + 6D^R \tag{1.4.4}$$

Equation 1.4.4 holds if translational and rotational diffusion are decoupled. For photocounts obeying Gaussian statistics, $g^{(1)}(q,t)$ is related to the intensity autocorrelation function $g^{(2)}(q,t)$ *via* the Siegert-relation [85, 90]:

$$g^{(2)}(q,t) = 1 + \mathcal{B}\left|g^{(1)}(q,t)\right|^2 \tag{1.4.5}$$

where $\mathcal{B}$ depends on the experimental setup.
If more than two relaxation processes contribute to the autocorrelation function, for example internal motions or little contaminations due to dust, the relaxation rate distribution function $G(\Gamma)$ can be calculated from $g^{(1)}(q,t)$ by inverse Laplace-transformation. In this thesis, the algorithm developed by Provencher was used for the determination of $G(\Gamma)$ [91, 92]:

$$g^{(1)}(q,t) = \int_{\Gamma=0}^{\Gamma=\infty} G(\Gamma) exp\left(-\Gamma t\right) d\Gamma \tag{1.4.6}$$

DDLS was applied to investigate the structure and the dynamics of various anisotropic objects in solution [22, 76, 88, 93, 94]. Piazza *et al.* investigated partial crystalline spherical and ellipsoidal latex particles under index-matching conditions [21, 95]. To calculate the shape parameters from the diffusion coefficients D^T and D^R, analytical expressions for simple geometries [96–98] and numerical solutions [99, 100] are available.
Because it is used in many areas of condensed matter physics and materials research, the conventional DLS/DDLS setup was further improved by using multiangle and multicorrelator instruments [94, 101]. While most studies analyzed the light measured far away from the sample (far field detection), recent developments focus on the detection of the light intensity near to the sample. According to Brogioli *et al.*, stray light problems can be eliminated in a DDLS experiment with near field detection, and a high number of statistical samples can be quickly obtained for a wide range of wave vectors in one measurement [102]. Very recently, Potenza *et al.* described a DDLS experiment which is based on a confocal, zero scattering angle heterodyne setup. As the signal comes almost entirely from particles in a small focal region, muliple-scattering contributions are effectively suppressed even in samples with large turbidity [77].

1.5 Objectives of this Thesis

The underlying topic of this dissertation is the fabrication of anisotropic colloidal particles and the study of their dynamics in solution by means of depolarized dynamic light scattering. For a general understanding of the counterion dynamics in a polyelectrolyte brush, the first contribution of the thesis aims to investigate both the hydrodynamic and the electrodynamic properties of spherical polyelectrolyte brushes (SPB) in an electric field. The influence of trivalent counterions on the elongation of the polyelectrolyte chains as well as the particle charge is characterized by measuring the electrophoretic mobility. This quantity is closely related to the particle potential ζ (see Chapter 2.1 and 3.1). The reliability of the experimental values ζ will be tested by a quantitative comparison with a theory, which takes into account the electrostatic, polymer and entropic contributions of the counter- and coions. The work aims to investigate properties of a brush layer which are of general importance for polyelectrolyte brushes.

The following studies are centered on i) modifying preparation strategies, which were established for spherical core-shell systems, to obtain dumbbell-shaped structures and on ii) investigating their translational and rotational diffusion in the highly diluted state. For the synthesis of the two dumbbell-shaped core-shell systems, two emulsion polymerization techniques are to be explored, namely photoemulsion- and seeded particle emulsion polymerization. The first technique will be utilized to generate a dense shell of polyelectrolyte chains on the preformed core particles to give dumbbell-shaped polyelectrolyte brushes (DPB, see Chapter 2.2 and 3.2). The second technique facilitates a crosslinked thermoresponsive polymer network (dumbbell-shaped core-shell microgel, DMP, see Chapter 2.3 and 3.3). As the novel colloids are either stimuli-responsive to the ionic strength (DPB) or the temperature (DMP), the influence of this parameters on the dynamics in solution will be quantified. The rotational and the translational diffusion coefficients will be used to determine the size parameters of the particles by hydrodynamic modeling. In addition to polarized (DLS) and depolarized dynamic light scattering (DDLS), field emission scanning electron microcsopy (FESEM), cryogenic-transmission electron microscopy (cryo-TEM), TEM and scanning force microscopy (SFM) serve as direct imaging techniques to clarify the particle structure.

To explore the fundamental role of particle geometry on translational and rotational diffusion, different highly defined colloidal clusters were chosen as ideal model systems. The results of the DDLS experiments are to be carefully compared with results from FESEM imaging and theoretical calculations using the shell model (Chapter 2.4 and 3.4).

1.6 References

[1] Perro, A.; Ravaine, S.; Duguet, E. *Actualite Chimique* **2010,** *337,* 14.

[2] Johnson, P. M.; van Kats, C. M.; van Blaaderen, A. *Langmuir* **2005,** *21,* 11510.

[3] Reculusa, S.; Poncet-Legrand, C.; Perro, A.; Duguet, E.; Bourgeat-Lami, E.; Mingotaud, C.; Ravaine, S. *Chem. Mater.* **2005,** *17,* 3338.

[4] Nie, Z.; Li, W.; Seo, M.; Xu, S.; Kumacheva, E. *J. Am. Chem. Soc.* **2006,** *128,* 9408.

[5] Sacanna, S.; Rossi, L.; Wouterse, A.; Philipse, A. P. *J. Phys.: Condens. Matter* **2007,** *19,* 376108.

[6] Manoharan, V. N.; Elsesser, M. T.; Pine, D. J. *Science* **2003,** *301,* 483.

[7] Kim, J.-W.; Larsen, R. J.; Weitz, D. A. *Adv. Mater.* **2007,** *19,* 2005.

[8] Cho, Y.-S.; Yi, G.-R.; Kim, S.-H.; Jeon, S.-J.; Elsesser, M. T.; Yu, H., K.; Yang, S.-M.; Pine, D. J. *Chem. Mater.* **2007,** *19,* 3183.

[9] Wagner, C. S.; Lu, Y.; Wittemann, A. *Langmuir* **2008,** *24,* 12126.

[10] Kraft, D. J.; Vlug, W. S.; van Kats, C. M.; van Blaaderen, A.; Imhof, A.; Kegel, W. K. *J. Am. Chem. Soc.* **2009,** *131,* 1182.

[11] Myers, D. *Surfaces, Interfaces, and Colloids: Principles and Applications.;* Wiley-VCH: New York: 2nd. ed.; 1999.

[12] Kim, J. W.; Larsen, R. J.; Weitz, D. A. *J. Am. Chem. Soc.* **2006,** *128,* 14374.

[13] Hosein, I. D.; Ghebrebrhan, M.; Joannopoulos, J. D.; Liddell, C. M. *Langmuir* **2010,** *26,* 2151.

[14] Yang, S.-M.; Kim, S.-H.; Lim, J.-M.; Yi, G.-R. *J. Mater. Chem.* **2008,** *18,* 2177.

[15] Lee, S. H.; Gerbode, S. J.; John, B. S.; Wolfgang, A. K.; Escobedo, F. A.; Cohen, I.; Liddell, C. M. *J. Mater. Chem.* **2008,** *18,* 4912.

[16] Zhang, R.; Schweizer, K. S. *Phys. Rev. E* **2009,** *80,* 011502.

[17] Vega, M.; Monson, P. A. *J. Chem. Phys.* **1997,** *107,* 2696.

[18] Williams, S. R.; Philipse, A. P. *Phys. Rev. E.* **2003,** *67,* 051301.

[19] Mock, E. B.; Zukoski, C. F. *Langmuir* **2007,** *23,* 8760.

[20] Mock, E. B.; Zukoski, C. F. *J. Rheol.* **2007,** *51,* 541.

[21] Degiorgio, V.; Piazza, R.; Corti, M.; Stavans, J. *J. Chem. Soc. Faraday Trans.* **1991,** *87,* 431.

[22] Lehner, D.; Lindner, H.; Glatter, O. *Langmuir* **2000,** *16,* 1689.

[23] Rühe, J. *et al. Adv. Polym. Sci.* **2004,** *165,* 79.

[24] Xu, Y.; Bolisetty, S.; Drechsler, M.; Fang, B.; Yuan, J.; Harnau, L.; Ballauff, M.; Müller, A. H. E. *Soft Matter* **2009,** *5,* 379.

[25] Guo, X.; Weiss, A.; Ballauff, M. *Macromolecules* **1999,** *32,* 6043.

[26] Guo, X.; Ballauff, M. *Langmuir* **2000,** *16,* 8719.

[27] Guo, X.; Ballauff, M. *Phys. Rev. E.* **2001,** *64,* 051406.

[28] Ballauff, M.; Borisov, O. *Curr. Opin. Colloid Interf. Sci.* **2006,** *11,* 306.

[29] Lu, Y.; Wittemann, A.; Ballauff, M. *Macromol. Rapid Commun.* **2009,** *30,* 806.

[30] Ballauff, M. *Prog. Polym. Sci.* **2007,** *32,* 1135.

[31] Pincus, P. *Macromolecules* **1991,** *24,* 2912.

[32] Wittemann, A.; Drechsler, M.; Talmon, Y.; Ballauff, M. *J. Am. Chem. Soc.* **2005,** *127,* 9688.

[33] Samokhina, L.; Schrinner, M.; Ballauff, M. *Langmuir* **2007,** *23,* 3615.

[34] Fritz, G.; Schädler, V.; Willenbacher, N.; Wagner, N. J. *Langmuir* **2002,** *18,* 6381.

[35] Jayachandranan, K. N.; Takacs-Cox, A.; Brooks, D. E. *Macromolecules* **2002,** *35,* 4247.

[36] Zhang, M.; Liu, L.; Zaho, H.; Fu, G.; He, B. *J. Colloid Interf. Sci.* **2006,** *301,* 85.

[37] Hariharan, R.; Biver, C.; Mays, J. W.; Russel, W. B. *Macromolecules* **1999,** *31,* 7506.

[38] Jusufi, A.; Likos, C. N.; Ballauff, M. *Colloid Polym. Sci.* **2004,** *282,* 910.

[39] Mei, Y.; Lauterbach, K.; Hoffmann, M.; Borisov, O. V.; Ballauff, M. *Phys. Rev. Lett.* **2006,** *97,* 158301.

[40] Mei, Y.; Hoffmann, M.; Ballauff, M.; Jusufi, A. *Phys. Rev. E.* **2008,** *77,* 031805.

[41] Mei, Y.; Lu, Y.; Polzer, F.; Ballauff, M. *Chem. Mater.* **2007,** *19,* 1062.

[42] Schrinner, M.; Proch, S.; Mei, Y.; Kempe, R.; Miyajima, N.; Ballauff, M. *Adv. Mater.* **2008,** *20,* 1928.

[43] Lu, Y.; Hoffmann, M.; Sai Yelamanchili, R.; Terrenoire, A.; Schrinner, M.; Drechsler, M.; Möller, M.; Breu, J.; Ballauff, M. *Macromol. Chem. Phys.* **2009,** *210,* 377.

[44] Wittemann, A.; Haupt, B.; Ballauff, M. *Phys. Chem. Chem. Phys.* **2003,** *5,* 1671.

[45] Schneider, C.; Jusufi, A.; Farina, R.; Li, F.; Pincus, P.; Tirrell, M.; Ballauff, M. *Langmuir* **2008,** *24,* 10612.

[46] Das, B.; Guo, X.; Ballauff, M. *Progr. Colloid Polym. Sci.* **2002,** *121,* 34.

[47] Dingenouts, N.; Patel, M.; Rosenfeldt, S.; Pontoni, D.; Narayanan, T.; Ballauff, M. *Macromolecules* **2004,** *37,* 8152.

[48] de Robillard, Q.; Guo, X.; Ballauff, M. *Macromolecules* **2000,** *33,* 9109.

[49] Hill, R. J.; Saville, D. A.; Russel, W. B. *J. Colloid Interf. Sci* **2003,** *258,* 56.

[50] Duval, J. F. L.; Ohshima, H. *Langmuir* **2006,** *22,* 3533.

[51] Zimmermann, R.; Norde, W.; Stuart, M. A. C.; Werner, C. *Langmuir* **2005,** *21,* 5108.

[52] Sheu, H. R.; El-Aasser, M. S.; Vanderhoff, J. W. *J. Polymer Sci. A. Polym. Chem.* **1990,** *28,* 653.

[53] Mock, E. B.; De Bruyn, H.; Hawkett, B. S.; Gilbert, R. G.; Zukoski, C. F. *Langmuir* **2006,** *22,* 4037.

[54] Kim, J.-W.; Lee, D.; Shum, H. C.; Weitz, D. A. *Adv. Mater.* **2008,** *20,* 3239.

[55] Lu, W.; Chen, M.; Wu, L. *J. Colloid Interf. Sci.* **2008,** *328,* 98.

[56] Hosein, I. D.; John, B. S.; Lee, S. H.; Escobedo, F. A.; Liddell, C. M. *J. Mater. Chem.* **2009,** *19,* 344.

[57] Ho, C. C.; Keller, A.; Odell, J. A.; Ottewill, R. H. *Colloid Polym. Sci.* **1993,** *271,* 469.

[58] Champion, J. A.; Katare, Y. K.; Mitragotri, S. *Proc. Natl. Acad. Sci. U. S. A.* **2007,** *104,* 11901.

[59] Sacanna, S.; Rossi, L.; Kuipers, B. W. M.; Philipse, A. P. *Langmuir* **2006,** *22,* 1822.

[60] Hoare, T.; Pelton, R. *Curr. Opinion Colloid Interf. Sci.* **2008,** *13,* 413.

[61] Lyon, L. A.; Meng, Z.; Singh, N.; Sorrell, C. D.; John, A. S. *Chem. Soc. Rev.* **2009,** *38,* 865.

[62] Pelton, R. *Adv. Colloid Interf. Sci.* **2000,** *85,* 1.

[63] Su, S.; Ali, M. M.; Filipe, C. D. M.; Li, Y.; Pelton, R. *Biomacromolecules* **2008,** *9,* 935.

[64] Blackburn, W. H.; Dickerson, E. B.; Smith, M. H.; McDonald, J. F.; Lyon, L. A. *Bioconjugate Chem.* **2009,** *20,* 960.

[65] Kang, J.-H.; Moon, J. H.; Lee, S.-K.; Park, S.-G.; Jang, S.-G.; Yang, S.; Yang, S.-M. *Adv. Mater.* **2008,** *20,* 3061.

[66] Dingenouts, N.; Norhausen, C.; Ballauff, M. *Macromolecules* **1998,** *31,* 8912.

[67] Lu, Y.; Wittemann, A.; Ballauff, M.; Drechsler, M. *Macromol. Rapid Commun.* **2006,** *27,* 1137.

[68] Berndt, I.; Pedersen, J. S.; Lindner, P.; Richtering, W. *Langmuir* **2006,** *22,* 459.

[69] Hellweg, T.; Dewhurst, C. D.; Eimer, W.; Kratz, K. *Langmuir* **2004,** *20,* 4330.

[70] Karg, M.; Pastoriza-Santos, I.; Liz-Marzan, L. M.; Hellweg, T. *ChemPhysChem* **2006,** *7,* 2298.

[71] Lu, Y.; Yu, M.; Drechsler, M.; Ballauff, M. *Macromol. Symp.* **2007,** *254,* 97.

[72] Suzuki, D.; McGrath, J. G.; Kawaguchi, H.; Lyon, L. A. *J. Phys. Chem. C.* **2007,** *111,* 5667.

[73] Crassous, J. J.; Siebenbürger, M.; Ballauff, M.; Drechsler, M.; Henrich, O.; Fuchs, M. *Chem. Phys.* **2006,** *125,* 204906.

[74] Hellweg, T.; Kratz, K.; Pouget, S.; Eimer, W. *Colloids Surf. A.* **2002,** *202,* 223.

[75] Kratz, K.; Hellweg, T.; Eimer, W. *Polymer* **2001,** *42,* 6631.

[76] Bolisetty, S.; Hoffmann, M.; Lekkala, S.; Hellweg, Th.; Harnau, L.; Ballauff, M. *Macromolecules* **2009,** *42,* 1264.

[77] Potenza, M. A. C.; Sanvito, T.; Alaimo, M. D.; Degiorgio, V.; Giglio, M. *Eur. Phys. J. E* **2010,** *31,* 69.

[78] Lauga, E.; Brenner, M. P. *Phys. Rev. Lett.* **2004,** *93,* 238301.

[79] Van Blaaderen, A. *Science* **2003,** *301,* 470.

[80] Yi, G.-R.; Manoharan, V. N.; Michel, E.; Elsesser, M. T.; Yang, S.-M.; Pine, D. J. *Adv. Mater.* **2004,** *16,* 1204.

[81] Cho, Y.-S.; Yi, G.-R.; Kim, S.-H.; Pine, D. J.; Yang, S.-M. *Chem. Mater.* **2005,** *17,* 5006.

[82] Zerrouki, D.; Rotenberg, B.; Abramson, S.; Baudry, J.; Goubault, C.; Leal-Calderon, F.; Pine, D. J.; Bibette, J. *Langmuir* **2006,** *22,* 57.

[83] Anthony, S. M.; Kim, M.; Granick, S. *J. Chem. Phys.* **2008,** *129,* 244701.

[84] Kim, M.; Anthony, S. M.; Granick, S. *Soft Matter* **2009,** *5,* 81.

[85] Berne, B. J.; Pecora, R. *Dynamic Light Scattering: With Applications to Chemistry, Biology and Physics;* Dover: New York: 2000.

[86] Perro, A.; Duguet, E.; Lambert, O.; Taveau, J.-C.; Bourgeat-Lami, E.; Ravaine, S. *Angew. Chem. Int. Ed.* **2009,** *48,* 361.

[87] Kraft, D. J.; Groenewold, J.; Kegel, W. K. *Soft Matter* **2009,** *5,* 3823.

[88] Eimer, W.; Williamson, J. R.; Boxer, S. G.; Pecora, R. *Biochemistry* **1990,** *29,* 799.

[89] Pecora, R. *Annu. Rev. Biophys. Bioeng.* **1972,** *1,* 257.

[90] Flamberg, A.; Pecora, R. *J. Phys. Chem.* **1984,** *88,* 3026.

[91] Provencher, S. W. *Comp. Phys. Commun.* **1982,** *27,* 213.

[92] Provencher, S. W. *Comp. Phys. Commun.* **1982,** *27,* 229.

[93] Matsuoka, H.; Morikawa, H.; Yamaoka, H. *Colloids Surf. A.* **1996,** *109,* 137.

[94] Shetty, A. M.; Wilkins, G. M. H.; Nanda, J.; Solomon, M. J. *J. Phys. Chem. C.* **2009,** *113,* 7129.

[95] Piazza, R.; Stavans, J.; Bellini, T.; Degiorgio, V. *Optics Commun.* **1989,** *73,* 263.

[96] Perrin, F. *J. Phys. Rad.* **1934,** *5,* 497.

[97] Perrin, F. *J. Phys. Rad.* **1936,** *7,* 1.

[98] Tirado, M. M.; Martinez, C. L.; Garcia de la Torre, J. *J. Chem. Phys.* **1984,** *81,* 2047.

[99] Carrasco, B.; Garcia de la Torre, J. *Biophys. J.* **1999,** *76,* 3044.

[100] Garcia de la Torre, J.; Del Rio Echenique, G.; Ortega, A. *J. Phys. Chem. B* **2007,** *111,* 955.

[101] Bantchev, G. B.; Russo, P. S.; McCarley, R. L.; Hammer, R. P. *Rev. Sci. Instrum.* **2006,** *77,* 043902.

[102] Brogioli, D.; Salerno, D.; Cassina, V.; Sacanna, S.; Philipse, A. P.; Croccolo, F.; Mantegazza, F. *Opt. Express* **2009,** *17,* 1222.

Chapter 2

Overview of this Thesis - Results

In this dissertation four publications are enclosed, presented from Chapter 3.1 to 3.4. The general question how trivalent counterions affect the hydrodynamic and electrodynamic properties of polyelectrolyte brushes in an electric field is answered in Chapter 2.1 and 3.1. As the results contain information about the dynamics in a brush layer, this work is closely related to Chapter 2.2 and 3.2.
The synthesis and the DDLS analysis of the dumbbell-shaped polyelectrolyte brushes (DPB) are presented in Chapter 2.2 and 3.2 together with data for a spherical reference system (SPB) as a function of the ionic strength.
In the following, it will be shown that the concept of thermoresponsive spherical core-shell microgels can be extended to the dumbbell-shaped analogue (DMP). The dynamics of the DMP in solution were examined in the same manner as for the DPB by using DDLS but as a function of the temperature (see Chapter 2.3 and 3.3).
While the previous studies are addressed to the dynamics of anisotropic particles with relatively simple shapes (sphere, dumbbell), the last part of the thesis is focused on complex-shaped colloids. Highly defined clusters with one to four spherical building blocks were studied by DDLS in combination with FESEM imaging and hydrodynamic modeling to understand the relation between the particle morphology and the diffusion coefficients for translational and rotational motion (see Chapter 2.4 and 3.4).
The main results of the four publications enclosed in this dissertation are summarized in Chapter 2.1 to 2.4. Manuscripts which have been additionally published within the scope of this disseration are listed on page 24.

The following publications and manuscripts are enclosed in this dissertation:

- **"Surface potential of spherical polyelectrolyte brushes in the presence of trivalent counterions"**
  Hoffmann, M.; Jusufi, A.; Schneider, C.; Ballauff, M. *J. Colloid Interf. Sci.* **2009**, *338*, 566-572.

- **"Dumbbell-Shaped Polyelectrolyte Brushes Studied by Depolarized Dynamic Light Scattering"**
  Hoffmann, M.; Lu, Y.; Schrinner, M.; Ballauff, M.; Harnau, L. *J. Phys. Chem. B.* **2008**, *112*, 13843-14850.

- **"Thermoresponsive Colloidal Molecules"**
  Hoffmann, M.; Siebenbürger, M.; Harnau, L.; Hund, M.; Hanske, C.; Lu, Y.; Wagner, C. S.; Drechsler, M.; Ballauff, M. *Soft Matter* **2010**, *6*, 1125-1128.

- **"3D Brownian Diffusion of Submicron-Sized Particle Clusters"**
  Hoffmann, M.; Wagner, C. S.; Harnau, L.; Wittemann, A. *ACS Nano*, **2009**, *3*, 3326-3334.

The following publications have been additionally published within the scope of this dissertation:

- **"Spherical polyelectrolyte brushes in the presence of multivalent counterions: The effect of fluctuations and correlations as determined by molecular dynamics simulations"**
  Mei, Y.; **Hoffmann, M.**; Ballauff, M.; Jusufi, A. *Phys. Rev. E* **2008**, *77*, 031805.

- **"Well-defined crystalline TiO2-Nanoparticles Generated and Immobilized on a Colloidal Nanoreactor"**
  Lu, Y.; **Hoffmann, M.**; Yelamanchili, R. S.; Terrenoire, A.; Schrinner, M.; Drechsler, M.; Möller, M. W.; Breu, J.; Ballauff, M. *Macromol. Chem. Phys.* **2009**, *210*, 377-386.

- **"Coupling of Rotational Motion with Shape Fluctuations of Core-Shell Microgels Having Tunable Softness"**
  Bolisetty, S.; **Hoffmann, M.**; Hellweg, T.; Harnau, L.; Ballauff, M. *Macromolecules*, **2009**, *42*, 1264-1269.

2.1 Surface Potential of SPB in the Presence of Trivalent Counterions

For the first time, this study presents experimental data for the surface potential ζ of fully characterized SPB in the presence of trivalent salt together with variational free energy calculations Ψ_{theo}. Earlier work made use of electrokinetic measurements to elucidate the unknown structure of polyelectrolyte layers appended to a substrate [1–3]. Our approach freshly reviews the problem by investigating the electrophoretic mobility of a model system, where the static structure and the thermodynamics are well-established [4] (see Chapter 1.1). The basic idea is to assume an identical shear plane for both diffusion (R_h) and the electrophoretic mobility (see Figure 2.1.1). The results are relevant to polyelectrolyte brushes in general.

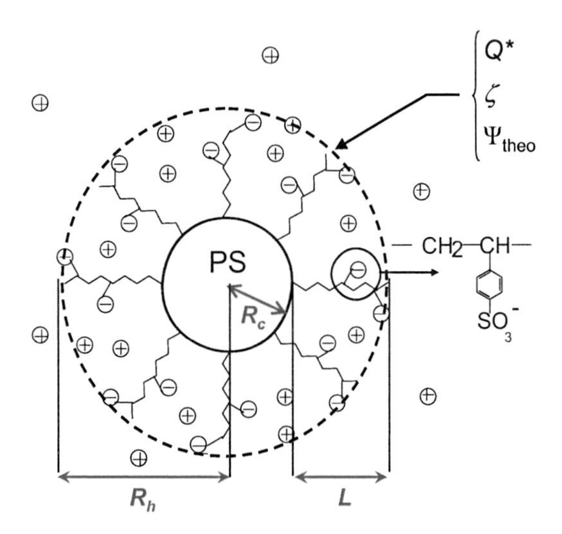

Figure 2.1.1: The experimental ζ-potential and the calculated surface potential Ψ_{theo} in this model are ascribed to the same hydrodynamic shear plane of the particle located at a distance R_h from the center of the core (dashed line). Immersing the SPB in solutions of $EuCl_3$ leads to an exchange of Na^+ with Eu^{3+} counterions which become immobilized in the brush. The free counterions outside the SPB give rise to the effective particle charge Q^* at the shear plane. Reprinted from [5], Copyright 2009, with permission from Elsevier.

The theory of O'Brien and White served to determine the particle potential ζ from the electrophoretic mobilities μ and the hydrodynamic radii R_h [6]. To apply this theory, the SPB were regarded as compact colloids. In order to simplify the calculation of the

equilibrium ionic strength, it was assumed that any Na^+ counterion in the SPB is replaced by the equivalent number of Eu^{3+} ions immediately. The calculation of Ψ_{theo} from the effective particle charge Q^* followed the lines given in refs. [4, 7].
Figure 2.1.2 shows that the surface potentials ζ and Ψ_{theo} agree at least semiquantitatively in the whole range of the ionic strength $I(EuCl_3)$. The steep decrease of the potential at $I(EuCl_3) = 2 \times 10^{-5}$ mol/L was assigned to a partial neutralization of charges by trivalent counterions accompanied with a collapse transition of the brush layer.

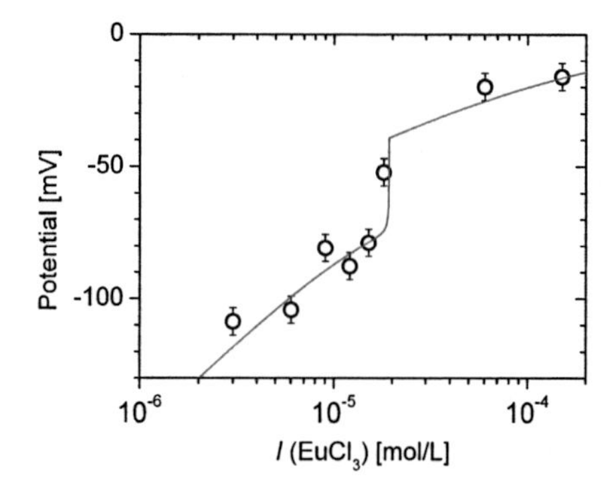

Figure 2.1.2: Surface potential Ψ_{theo} for the SPB in $EuCl_3$ solutions as determined by variational free energy calculations (line) and the ζ-potential derived from the experimental electrophoretic mobilities (circles). Reprinted from [5], Copyright 2009, with permission from Elsevier.

The findings indicate that the SPB behave as compact colloids rather than soft ones in the presence of trivalent counterions. In this case, electroosmotic flows [8] inside the brush caused by the external electric field on the counterions should be neglegible. This points to a strong suppression of electroosmotic flows inside the brush layer due to the strong correlation of the counterions to the polyelectrolyte chains. In general, this was the first experimental study which casted light on the dynamics of multivalent counterions in a polyelectrolyte brush (compare ref. [9]).

2.2 Dumbbell-Shaped Polyelectrolyte Brushes Studied by DDLS

While the previous study clarified the dynamics of trivalent counterions in a brush layer, this work is devoted to the internal and the overall dynamics of polyelectrolyte brushes with different morphologies. Stable anisotropic colloidal core particles were synthesized and modified by grafted polymer for the first time. These so-called dumbbell-shaped polyelectrolyte brushes (DPB) have a core made of poly(methyl methacrylate) (PMMA) and poly(styrene) (PS) from which a dense brush layer of poly(styrene sulfonate) chains (PSS) was grafted. The particle morphology was controlled by using relatively hydrophilic PMMA seed particles and styrene as a hydrophobic monomer (thermodynamic control) [10]. The addition of monomer under starved conditions at 60 °C minimized the diffusion of unreacted monomer into the seed particle (kinetic control) [11] (see Scheme 2.2.1).

Scheme 2.2.1: Scheme for the preparation of dumbbell-shaped polyelectrolyte brushes (DPB)[a]

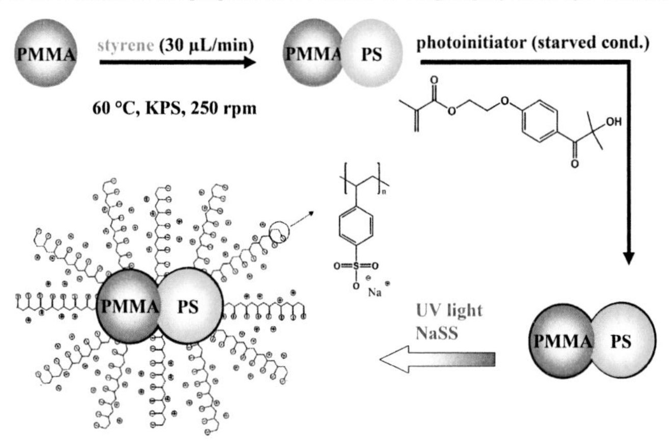

[a] First, PMMA particles are prepared by conventional emulsion polymerization. The dumbbell morphology is formed when adding styrene under starved conditions ($30\,\mu$L·min^{-1}). In the next step, these core particles are covered with a thin layer of photoinitiator HMEM. In the last step, the shell of the polyelectrolyte brushes is formed upon irradiation with UV light in a photoemulsion polymerization. Reprinted with permission from [12]. Copyright 2008 American Chemical Society.

DDLS was the method of choice to determine the translational D^T and the rotational diffusion coefficient D^R for the stimuli-responsive DPB, as the DPB were both

shape and optical anisotropic [13]. Measurements at different ionic strength (I =1 and 100 mM) demonstrated that the overall size and dynamics of the DPB depended markedly on the brush layer thickness.
As a central finding of this work, the relaxation rates of the intensity autocorrelation functions gave access to an new relaxation process at higher frequencies. This could be assigned to a collective relaxation of the brush layer rather than of individual polymer chains. Both the collective and the rotational relaxation were observed for spherical polyelectrolyte brushes as well. (Figure 2.2.1). Thus, there is a strong indication that the collective relaxation mode is the dynamic counterpart of the static scattering intensities measured earlier in the case of SPB particles which is due to spatial fluctuations of the grafted chains [9, 14].

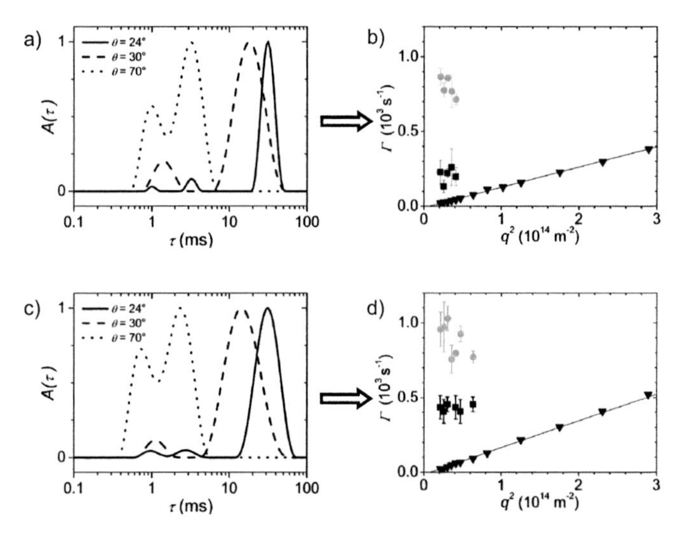

Figure 2.2.1: DDLS-relaxation time distributions (CONTIN-plots) calculated from the intensity autocorrelation functions for the DPB particles in a 1 mM (a) and a 100 mM NaCl solution (c). For the scattering angle $\theta = 24°$, three distinct relaxation modes can be resolved (solid lines) which are connected with the translational motion (slow mode, $\approx 30\,\text{ms}$), the particle rotation (fast mode, $\approx 2-3\,\text{ms}$), and a collective relaxation process (fast mode, $\approx 1\,\text{ms}$). For $\theta = 30°$, the two faster modes overlap and cannot be resolved (dashed lines). (b,d) The corresponding relaxation rates Γ are plotted as a function of the square of the scattering vector (q^2). Here the translational motion (slow mode) is assigned by diamonds (▼), the rotational relaxation around the particle minor axis together with translational motion by squares (■), and the collective relaxation by circles (●). Reprinted with permission from [12]. Copyright 2008 American Chemical Society.

2.3 Thermoresponsive Colloidal Molecules

Thermoresponsive colloidal molecules (DMP) of approximately 250 nm in size were fabricated and characterized by a combination of polarized (DLS) and depolarized dynamic light scattering (DDLS), electron microscopy (cryo-TEM, FESEM) and scanning force microscopy (SFM). The DMP present the first example of non-rigid colloidal molecules, which facilitate the variation of the shape and the aspect ratio between 1.4. and 1.6 simply by changing the temperature. Therefore, no chemical modification is needed to change the particle shape once synthesized compared to previous studies [15–17].
To obtain the DMP, the synthetic approach originally developed for spherical core-shell microgels [18] had to be modified. The preparation of the core particles (see Chapter 2.2 and 3.2) had to be carried out without N-isopropylacrylamide to avoid an influence on the resulting core particle morphology (see Figure 2.3.1a).
The charge induced 2-dimensional self-assembly of the novel colloids on a Si and a glass substrate was verified by FESEM and SFM as shown in Figure 2.3.1b and d. Cryo-TEM proved the dumbbell-shaped particle morphology *in situ* (see Figure 2.3.1c).

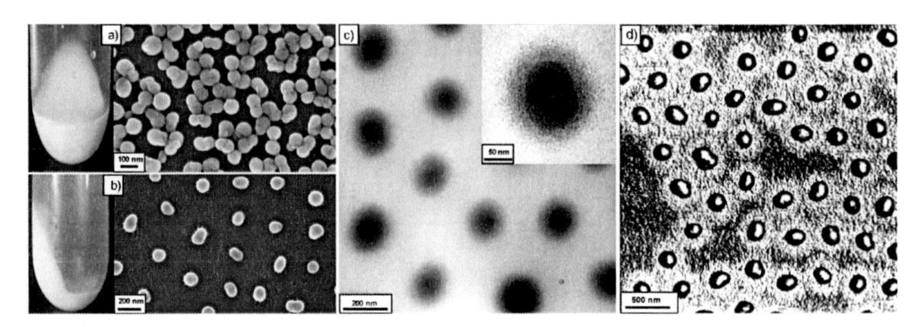

Figure 2.3.1: FESEM micrograph of the a) dumbbell-shaped core particles and the b) thermoresponsive core-shell particles (DMP). The collapsed PNIPA shell spreads over a Si wafer. After centrifugation of the suspensions, iridescence can be observed. c) Cryo-TEM micrographs of the core-shell microgel particles. The shell thickness can be estimated as (51.5 ± 5.9) nm for 23 °C. d) SFM phase image of the DMP particles deposited from a 0.003 wt % solution on a glass slide in air (phase angle $0 - 20°$). [19] - Reproduced by permission of The Royal Society of Chemistry.

DDLS was employed to obtain the translational and the rotational diffusion coefficients D^T and D^R for the DMP, respectively. For temperatures between 14.8 and 36.8 °C, the acceleration of the particle dynamics could be ascribed to the shrinking of the thermoresponsive poly(N-isopropylacrylamide) (PNIPA) layer.
To model the hydrodynamics of the DMP, this work makes use of more sophisticated analytical tools compared to the previous study [12]. Applying the

shell model [20] facilitated the quantitative calculation of the PNIPA layer thickness L_h (see Figure 2.3.2). The diffusion coefficients of the core particles without the PNIPA layer were used to obtain the center-to-center distance l and the radius of one constituent sphere R. The shell thickness L_h was varied in such way that the theoretical diffusion coefficients D^T_{theo} and D^R_{theo} matched best the experimental values D^T and D^R for each temperature. The value of about 60 nm for L_h at 23 °C agreed well with microscopic evidence (see Figure 2.3.1c). This is central to apply the synthetic approach independent of the core particle geometry. The shell thickness of the PNIPA layer was the same for the DMP particles and a spherical reference system within the experimental error (see Figure 2.3.2).

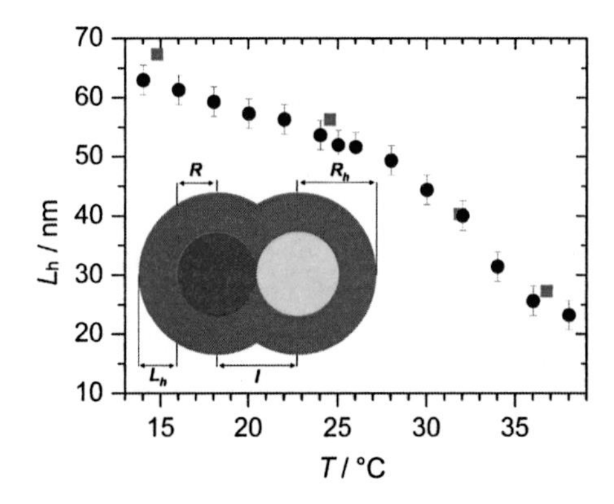

Figure 2.3.2: Shell thickness L_h as a function of the temperature as determined for the DMP particles with the shell model (■) together with the data for a spherical reference system calculated using the Stokes-Einstein equation and the core particle radius (●). Inset: Sketch of the shell model used for the calculation of the shell thickness $L_h = R_h - R$. The shell model considers the particle surface composed of small, spherical friction elements under stick-boundary conditions and takes into account an interpenetration of two spheres with radius R_h and a center-to-center distance l. [19] - Reproduced by permission of The Royal Society of Chemistry.

2.4 Hydrodynamic Description of Well-Defined Colloidal Clusters

The rotational diffusion, which was measured for the previous model systems, was due to the relaxation of the main particle axis of the particles. To study the motion of objects with complex shape, colloidal clusters with well-defined configurations [21] were chosen as ideal model systems. As the clusters with a size below 300 nm were stable against sedimentation, the true 3-dimensional diffusion purely to Brownian motion was probed. Hence, this work overcomes previous limitations with micron-sized particles near interfaces as wall effects could be excluded [22, 23]. For this work, polarized (DLS) and depolarized dynamic light scattering (DDLS) were applied to measure the translational diffusion D^T and the rotational relaxations D^R of the colloidal clusters. As colloidal particles with a complex shape exhibit different rotational relaxations around all axes of symmetry, which may not all give rise to the DDLS signal, the precise predictions of the shell model [20] served to interpret the experimental data.
Figure 2.4.1 shows the experimental diffusion coefficients D^T and D^R as a function of the number N of building blocks of the clusters. The decay of D^T widely follows from the increase in the mean radius with the number N of the clusters. However, for D^R a marked drop was observed when going from $N = 1$ to $N = 2$, while the value of D^R was almost the same when going from $N = 2$ to $N = 3$. It was found that the rotational and the translational diffusion are decoupled.

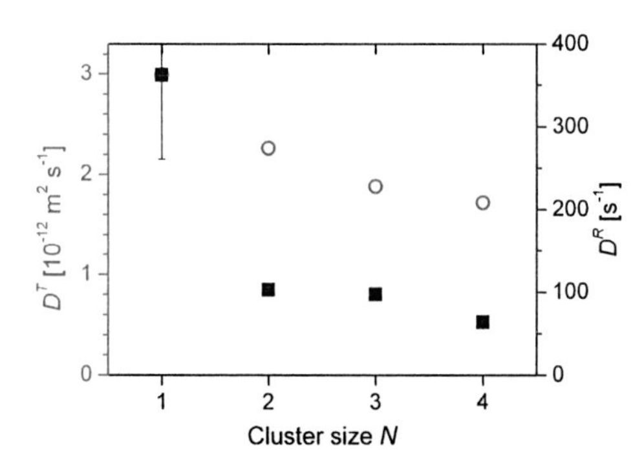

Figure 2.4.1: Translational D^T (○) and rotational diffusion coefficients D^R (■) of the particle clusters as a function of the number of building blocks N (single particles, $N = 1$; particle doublets, $N = 2$; triplets, $N = 3$; tetrahedrons, $N = 4$). Reprinted with permission from [24]. Copyright 2009 American Chemical Society.

Central to the understanding of the rotational dynamics of the clusters in solution was to assign the rotational relaxation measured by DDLS to the corresponding axis of symmetry in the cluster (see Figure 2.4.2). For the calculations with the shell model, the size of the clusters as determined from FESEM images was used. The experimental diffusion coefficients were in excellent agreement with theoretical predictions which assume stick-boundary conditions. As a comparison of the experimental D^R and the theoretical diffusion coefficients $D^R_\perp$ revealed, DDLS only probed the rotational relaxation of the clusters with $N = 2$ around the minor axis and in the case of $N = 3$ around the axis in the plane of the triplet. The relaxations around the main particle axes $D^R_\parallel$ could not be detected due to the rather small anisotropy of the spherical building blocks. It was found that the rotation of the tetrahedron cannot be assigned to a specific axis, as the diffusion of tetrahedrons resembles those of a sphere due to its low shape anisotropy. Despite of the spherical geometry of the singlets ($N = 1$) a DDLS signal was observed. We attributed this to the small optical anisotropy of the building blocks.

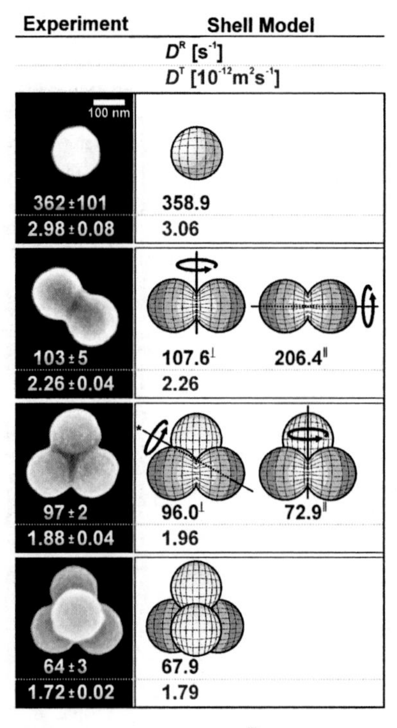

Figure 2.4.2: Comparison of the translational (D^T; lower value) and rotational ($D^R_\parallel$, $D^R_\perp$; upper value) diffusion coefficients as obtained by DLS and DDLS, respectively, together with the theoretical results using the shell model. For the particle doublets and triplets, the rotational diffusion coefficient $D^R_\perp$ perpendicular to the main symmetry axis is measured. In the left column, the particle clusters are oriented with their main body parallel to the plane of the figure. Reprinted with permission from [24]. Copyright 2009 American Chemical Society.

2.5 Individual Contributions to Joint Publications

The results presented in this thesis were obtained in cooperation with others and published as indicated below. In the following, the contributions of all the coauthors to the different publications are specified. The asterisk denotes the corresponding author.

Chapter 3.1

- This work is published in the JOURNAL OF COLLOID AND INTERFACE SCIENCE (**2009**, *338*, 566-572) under the title:

"Surface potential of spherical polyelectrolyte brushes in the presence of trivalent counterions"
by Martin Hoffmann, Arben Jusufi, Christian Schneider and Matthias Ballauff*

- I conducted all the analytical experiments, calculated the experimental potential from the mobilities, and wrote the publication.
- C. Schneider prepared the spherical polyelectrolyte brushes and contributed to scientific discussion.
- A. Jusufi performed the computer simulation of the spherical polyelectrolyte brushes and contributed to the scientific discussion and correcting the manuscript.
- M. Ballauff was involved in scientific discussion and correcting the manuscript.

Chapter 3.2

- This work is published in the JOURNAL OF PHYSICAL CHEMISTRY B (**2008**, *112*, 14843-14850) under the title:

"Dumbbell-Shaped Polyelectrolyte Brushes Studied by Depolarized Dynamic Light Scattering"
by Martin Hoffmann, Yan Lu, Marc Schrinner, Matthias Ballauff* and Ludger Harnau

- I conducted the synthesis of the brush particles, most of the analytical experiments and wrote the publication. Exceptions are stated in the following:
- M. Schrinner performed the TEM and cryo-TEM measurements.
- Y. Lu performed the FESEM measurements and was involved in scientific discussion.
- L. Harnau and M. Ballauff were involved in scientific discussion and correcting the manuscript.

Chapter 3.3

- This work is published in SOFT MATTER (**2010**, *6*, 1125-1128) under the title:

"Thermoresponsive Colloidal Molecules"
by Martin Hoffmann, Miriam Siebenbürger, Ludger Harnau, Markus Hund, Christoph Hanske, Yan Lu, Claudia Simone Wagner, Markus Drechsler and Matthias Ballauff*

- I synthesized the anisotropic core shell colloids, performed the dynamic light scattering experiments (DLS, DDLS) together with the data analysis and wrote the publication.
- M. Siebenbürger conducted the DLS measurements of the spherical core-shell microgel reference system and contributed to scientific discussion.
- L. Harnau described the particle dynamics using the shell model and contributed to scientific discussion.
- M. Hund and C. Hanske performed the SFM measurements.
- Y. Lu contributed to scientific discussion.
- C. S. Wagner performed the FESEM measurements.
- M. Drechsler performed the cryo-TEM measurements.
- M. Ballauff contributed to scientific discussion and correcting the manuscript.

Chapter 3.4

- This work is published in ACS Nano (**2009**, *3*, 3326-3334) under the title:

"3D Brownian Diffusion of Submicron-Sized Particle Clusters"
by Martin Hoffmann, Claudia Simone Wagner, Ludger Harnau and Alexander Wittemann*

- I performed all dynamic light scattering experiments (DLS, DDLS), the analysis of the scattering data and wrote parts of the publication.
- C. S. Wagner prepared the colloidal clusters and performed the FESEM measurements.
- L. Harnau described the cluster dynamics using different hydrodynamic models (spheroids; shell model), contributed to scientific discussion and writing the manuscript.
- A. Wittemann contributed to scientific discussion and writing the manuscript.

2.6 References

[1] Zimmermann, R.; Norde, W.; Stuart, M. A. C.; Werner, C. *Langmuir* **2005,** *21,* 5108.

[2] Duval, J. F. L.; Ohshima, H. *Langmuir* **2006,** *22,* 3533.

[3] Dukhin, S. S.; Zimmermann, R.; Werner, C. *J. Colloid Interf. Sci.* **2008,** *328,* 217.

[4] Jusufi, A.; Likos, C. N.; Ballauff, M. *Colloid Polym. Sci.* **2004,** *282,* 910.

[5] Hoffmann, M.; Jusufi, A.; Schneider, C.; Ballauff, M. *J. Colloid Interf. Sci.* **2009,** *338,* 566.

[6] O'Brien, R. W.; White, L. R. *J. Chem. Soc. Faraday Trans. II* **1978,** *74,* 1607.

[7] Schneider, C.; Jusufi, A.; Farina, R.; Li, F.; Pincus, P.; Tirrell, M.; Ballauff, M. *Langmuir* **2008,** *24,* 10612.

[8] Hill, R. J.; Saville, D. A.; Russel, W. B. *J. Colloid Interf. Sci* **2003,** *258,* 56.

[9] Dingenouts, N.; Patel, M.; Rosenfeldt, S.; Pontoni, D.; Narayanan, T.; Ballauff, M. *Macromolecules* **2004,** *37,* 8152.

[10] Chen, Y.-C.; Dimonie, V.; El-Aasser, M. S. *J. Appl. Polym. Sci.* **1991,** *42,* 1049.

[11] Stubbs, J. M.; Karlsson, O.; Jönsson, J. E.; Sundberg, E.; Durant, Y.; Sundberg, D. *Coll. Surf. A* **1999,** *153,* 255.

[12] Hoffmann, M.; Lu, Y.; Schrinner, M.; Ballauff, M.; Harnau, L. *J. Phys. Chem. B:* **2008,** *112,* 13843.

[13] Berne, B. J.; Pecora, R. *Dynamic Light Scattering: With Applications to Chemistry, Biology and Physics;* Dover: New York: 2000.

[14] de Robillard, Q.; Guo, X.; Ballauff, M. *Macromolecules* **2000,** *33,* 9109.

[15] Kraft, D. J.; Groenewold, J.; Kegel, W. K. *Soft Matter* **2009,** *5,* 3823.

[16] Kraft, D. J.; Vlug, W. S.; van Kats, C. M.; van Blaaderen, A.; Imhof, A.; Kegel, W. K. *J. Am. Chem. Soc.* **2009,** *131,* 1182.

[17] Perro, A.; Duguet, E.; Lambert, O.; Taveau, J.-C.; Bourgeat-Lami, E.; Ravaine, S. *Angew. Chem. Int. Ed.* **2009,** *48,* 361.

[18] Dingenouts, N.; Norhausen, C.; Ballauff, M. *J. Phys. Chem. B* **1998,** *31,* 8912.

[19] Hoffmann, M.; Siebenbürger, M.; Harnau, L.; Hund, M.; Hanske, C.; Lu, Y.; Wagner, C. S.; Drechsler, M.; Ballauff, M. *Soft Matter* **2010,** *6,* 1125.

[20] Garcia de la Torre, J.; Del Rio Echenique, G.; Ortega, A. *J. Phys. Chem. B* **2007,** *111,* 955.

[21] Wagner, C. S.; Lu, Y.; Wittemann, A. *Langmuir* **2008,** *24,* 12126.

[22] Anthony, S. M.; Kim, M.; Granick, S. *J. Chem. Phys.* **2008,** *129,* 244701.

[23] Kim, M.; Anthony, S. M.; Granick, S. *Soft Matter* **2009,** *5,* 81.

[24] Hoffmann, M.; Wagner, C. S.; Harnau, L.; Wittemann, A. *ACS Nano* **2009,** *3,* 3326.

Chapter 3

Publications of this Thesis

3.1 Surface Potential of Spherical Polyelectrolyte Brushes in the Presence of Trivalent Counterions

Martin Hoffmann[a], Arben Jusufi[b], Christian Schneider[a] and Matthias Ballauff[a*]

[a] *Physikalische Chemie I, University of Bayreuth, 95440 Bayreuth, Germany*
[b] *Laboratory for Research on the Structure of Matter (LRSM), University of Pennsylvania, Philadelphia, PA, 19104, U.S.A.*

*Corresponding author: Matthias.ballauff@uni-bayreuth.de

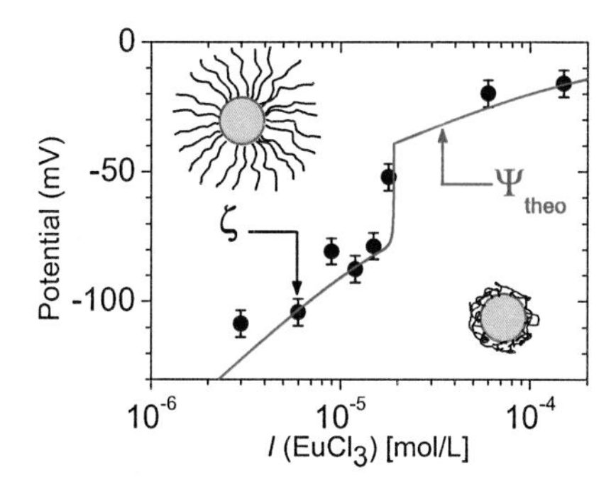

Published in the *Journal of Colloid and Interface Science* **2009**, *338*, 566-572.

3.1.1 Abstract

We consider the ζ-potential and the effective charge of spherical polyelectrolyte brushes (SPBs) in aqueous solution in the presence of trivalent europium ions. The SPB consists of a polystyrene core of ca. 250 nm diameter onto which long chains of the strong polyelectrolyte poly(styrene sulfonate) are grafted (contour length: 82 nm). At low concentrations of $EuCl_3$ the chains are stretched to nearly full length. If the concentration of the trivalent ions is raised, the surface layer of the polyelectrolyte chains collapses. The ζ-potential of the SPB is calculated from the electrophoretic mobilities measured at different concentrations of $EuCl_3$. At the collapse, ζ decreases by the partial neutralization of the charges by the trivalent ions. The experimental ζ-potential thus obtained agrees with the theoretical surface potential Ψ_{theo} calculated for the effective shear plane by a variational free energy model of the SPBs.

KEYWORDS: electrophoretic mobility, zeta-potential, brush, effective charge, dynamic light scattering, variational free energy calculation

3.1.2 Introduction

The electrostatic stabilization of colloids by surface charges is among the best-studied problems in colloid science [1,2]. It is by now well-understood that the high electric field resulting from the bare charge Q_b leads to partial condensation of the counterions onto the surface and the electrostatic stability will be provided only by the effective charge Q^*. If Q^* on the surface is known, the corresponding surface potential Ψ is accessible and the repulsive potential between two colloidal spheres can be calculated quantitatively. The effective charge will also determine the electrophoretic mobility μ of the particles in suspension as well as the ζ-potential in an electric field. As a convention, the ζ-potential is defined as the electrostatic potential at a shear plane [3-5]. The ζ-potential has been used to assess the strength of electrostatic repulsion and thus the colloidal stability of the suspensions [1,2].
Early studies of Hückel [6], Smoluchowski [7] and Henry [8] provide analytical expressions for $\mu(\zeta)$ at low surface potentials but neglect the relaxation effect of the electric double layer. For highly charged rigid particles in mono- or multivalent salt, the theory of O'Brien and White (OW) [9] can be applied for the conversion of the electrophoretic mobility into the ζ- potential. This theory includes double layer relaxation which originates from the polarization of ions within the diffuse layer around the particle by the applied electrical field. A careful comparison of theory and experiment demonstrates that the OW model provides a good description for highly charged sulfonated polystyrene spheres [10,11]. The obvious practical importance of the ζ-potential has led to a large number of theoretical [12-15] and experimental studies [5,16]. Experimental data obtained with polystyrene sulfate colloids in the presence of divalent ions are discussed in Ref. [17]. The relevance of ion size correlations especially for di- and trivalent ions on the mobility of sulfonated latexes is demonstrated in Ref. [18]. It is fair to state that the relation of the electrophoretic mobility and of the ζ-potential to Q^* and the colloidal stability of charged spherical colloids seems to be well understood by now. A comprehensive review of this problem has been given recently [5].
However, an entirely different situation arises if long polyelectrolyte chains are appended to the surface of the spherical particles. These "soft particles" can be made either by adsorption from the solution, by grafting to the surface [5,19-22], or by affixing a polymeric network to solid core particles [23,24]. The effect of particle coating with polyelectrolytes complicates the calculation of the electrophoretic mobility in a significant manner: First of all, the polymer layer may immobilize or release a certain fraction of its counterions depending on the ionic strength in the system. In this way the effective charge and the thickness of the coating become a function of the concentration and valency of the added salt as recently discussed by Hill, Saville, and Russel [19,20]. Moreover, the hydrodynamic permeability has a profound influence on the mobility [25,26]. If counterions in the soft layer exhibit a non-zero mobility they can induce an electroosmotic flow in the presence of an external electric field. Thus, the soft layer becomes permeable to the solvent.

The resulting fluid flow is impeded by the hydrodynamic drag exerted by the polymer segments [20]. The extent to which solvent convection develops within the soft layer is given by the penetration length $\lambda_0^{-1} = (\eta_W/K)^{1/2}$ (η_W dynamic viscosity of water, K friction coefficient exerted by the polymer segments) [27] or the Brinkman screening length $l_B = (n_S 6\pi a_S F_S)^{-1/2}$ (a_S Stokes radius of a polymer segment, n_S polymer segment density, F_S dimensionless drag coefficient for random sphere configuration [19,28]) as expressed by Hill [19]. Duval and Ohshima derived a theory for soft particles [15,27]. In this model, the electrophoretic mobility is calculated for a given distribution of polymer segments at diffuse soft interfaces. As shown by Dukhin *et al.*, the internal structure of polyelectrolyte layers (PL) may be inferred from surface conductivity measurements in order to calculate the Donnan potential and the intrinsic charge [29-32]. Furthermore, measurements of the streaming potential were conducted to investigate the electrokinetic properties of thin films of crosslinked charged polyacrylamide copolymer gels [33] or of planar poly(acrylic acid) brushes [34].

Much of the previous work has been done in order to use electrophoretic measurements to elucidate the structure of polymeric surface layers. The present study aims at a different approach of this problem by investigating the electrophoretic mobility of a model system where the static structure and the thermodynamics is well-established. Figure 3.1.1 displays this model system termed spherical polyelectrolyte brush (SPB; [22]). The particle consists of a solid polymer core with known size (R_c) onto which long linear chains of poly(styrene sulfonate) of length L are densely grafted (σ). Hence, the appended chains interact strongly (brush-limit; cf. Ref. [35-37]) and most of the counterions are immobilized within the polyelectrolyte layer [38-40]. For salt-free solution the osmotic pressure of the confined counterions leads to a strong stretching of the chains (osmotic brush)[22,35,36,41-43]. If the ionic strength is raised by adding monovalent salt, the brush layer shrinks gradually until the limit of the salted brush is reached [22,35-37]. A similar effect has been observed in studies of charged microgels [23]. However, trivalent ions lead to a much more pronounced shrinking at the same ionic strength since the osmotic pressure within the brush layer is strongly reduced [43,44]. This leads to a collapse of the brush layer at a given concentration of trivalent ions that can be followed precisely by dynamic light scattering (DLS; see Ref. [43]). The SPBs may be used to remove multivalent ions in the treatment of radioactive waste water [45]. However, such applications require the precise control of particle stability.

Here we present the analysis of the electrophoretic mobility μ and ζ-potential of spherical polyelectrolyte brushes in the presence of trivalent europium counterions. The present study has been prompted by the success of the theoretical description of the SPB in presence of trivalent ions [43,44]: The reduction of the height L of the surface layer as the function of salt concentration can be quantitatively understood when considering the balance of the osmotic pressure of the counterions within the brush layer and the retracting force of the polymer chains [43,44]. Thus, the thickness L of the surface layer

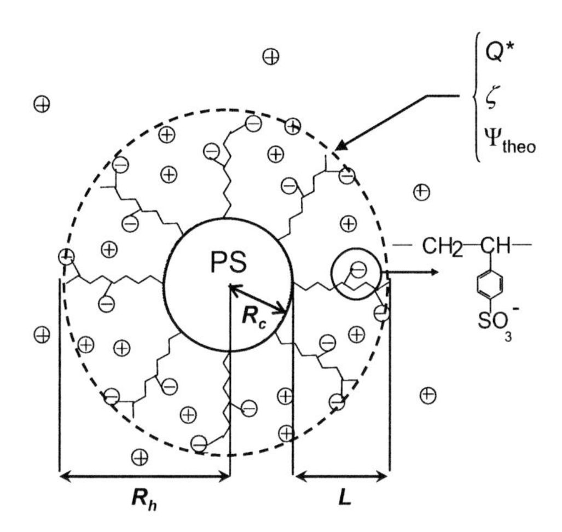

Figure 3.1.1: Schematic representation of the spherical polyelectrolyte brush under consideration. The core of radius R_c consists of polystyrene and a thin layer of photoiniiator, from which polyelectrolyte chains made of sodium styrene sulfonate are densely grafted (σ). L denotes the thickness of the brush shell, R_h the hydrodynamic radius. The effective particle charge Q^*, the experimental ζ-potential and the calculated surface potential Ψ_{theo} in this model are ascribed to the hydrodynamic shear plane of the particle located at a distance R_h from the center of the core (dashed line).

(see Figure 3.1.1) can be modeled as the function of the concentration of the added trivalent ions. An important ingredient of this theory is the partial neutralization of the SPB by the trivalent counterions that are fully condensed onto the polyelectrolyte chains [44]. These ions do not have translational entropy and do not contribute to the osmotic pressure inside the brush [44]. The loss of osmotic pressure inside the layer of polyelectrolyte chains is followed by a marked collapse of the surface layer [43-44]. Concomitantly, the colloidal stability of the particles is expected to decrease very much which is found indeed [46]. As a consequence of this, the partial neutralization of the chains by the trivalent ions should be followed by a strong decrease of the effective charge Q^* and the ζ-potential. Up to now, however, no experimental data are available for the ζ-potential and Q^* of these model SPB in presence of trivalent counterions. The problem at hand can be divided in two parts:

i) In the first step the hydrodynamic radius R_h and the electrophoretic mobility μ are measured for the SPBs in various concentrations of $EuCl_3$ [47] (see Section 3.1.5.1). From the electrophoretic mobilities, the $\zeta-$ potential can be deduced using the theory of O'Brien and White [9]. For this we assume that the effective shear plane [5] coincides

for both diffusion as well as for the electrophoretic mobility (see Section 3.1.5.2).
ii) In a second step the surface potential Ψ_{theo} at the shear plane defined by R_h is calculated from theory without adjustable parameters. Here, R_h is obtained from the theory delineated in Ref. [48]. The same approach gives the effective charge Q^*. The effective charge can be related to the effective shear plane and used for the calculation of the surface potential Ψ_{theo} (see Section 3.1.4).
A comparison of Ψ_{theo} thus obtained with the experimental ζ may then serve for a test of the underlying assumptions (see Section 3.1.5.3).

3.1.3 Materials and Methods

3.1.3.1. Materials

The core particles as the precursor for the SPB were synthesized in an emulsifier-free emulsion polymerization of styrene together with the functional monomer sodium (styrene sulfonate) (NaSS) using a solution of $NaHCO_3$, Na_2SO_3 and $K_2S_2O_8$ (KPS) as redox-initiator as described in Ref. [49]. The synthesis of the photoinitiator 2-[p-(2-Hydroxy-2-methylpropiophenone)] - ethylenglycol methacrylat (HMEM) and the photoemulsion polymerization reaction are described in Ref. [50,51].
The hydrodynamic radius of the core particles R_c in 10 mM NaCl solution was (126 ± 2) nm and the thickness of the poly(styrene sulfonate) shell (PSS) on the core particles in water (74 ± 3) nm (DLS). To determine the contour length L_c of the strong polyelectrolyte chains, the brush layer was cleaved off using a strong base [50]. Separation of chains from the remaining core particles and salt was achieved through extensive ultrafiltration using membranes of different pore sizes (200 nm and 5000 Da cut off membrane; nitrocellulose, Schleicher & Schuell). The freeze dried powder was analyzed by Size Exclusion Chromatography calibrated with linear PSS-standards (Polymer Standard Service, Germany). The corresponding contour length of the brush was calculated to be L_c=(82 ± 6) nm, where we assumed that the size of one monomer unit of NaSS is 0.25 nm. The number of negatively charged NaSS units per SPB particle was $N(\text{NaSS}) = (1840000 \pm 644000)$ accompanied with a grafting density of the polyelectrolyte shell σ of (0.029 ± 0.010) PSS chains per nm^2 of the core particle surface.

3.1.3.2. Measurements of the electrophoretic mobility for the SPB in solutions of $EuCl_3$

The electrophoretic measurements were performed using a Zetasizer Nano ZS instrument (Malvern Instruments) under a scattering angle of 17° at 25 °C with folded capillary cells (Malvern Instruments). The inner walls of each capillary cells were carefully rinsed twice with 20 ml of pure ethanol, water (Millipore) and subsequently the salt solution of the same concentration as desired for the measurement later. For the sample preparation we immersed 0.0025 wt% of SPB particles in the salt solutions with ionic strength $I(EuCl_3)$ (ca. 3.04×10^{12} SPB particles/liter). The system was allowed to equilibrate for 10 minutes. For each salt concentration we measured and averaged 5 series, each being the result of 20 to 25 single runs.

3.1.3.3. Dynamic Light Scattering

The experiments were carried out with an ALV/DLS/DLS-5000 compact goniometer system (Peters) equipped with a He-Ne laser at a wavelength of 632.8 nm, an ALV-6010/160 External Multiple Tau Digital Correlator and a thermostat (Rotilabo ±0.1 °C). 0.0025 wt% solutions of SPB in different salt concentrations were filled into quartz cuvettes (Hellma) which were immersed in an index matching *cis*-decaline bath. We performed 5 single runs for each sample at a scattering angle of 90° for 180 seconds resulting in count rates around 150 kHz. The standard cumulant analysis was applied to calculate the mean relaxation frequency $\Gamma = D^T|\vec{q}|^2$, where the translational diffusion coefficient D^T is connected with the hydrodynamic radius R_h by the Stokes-Einstein-equation through $R_h = kT/(6\pi\eta D^T)$, where $|\vec{q}| = 4\pi \sin\left(\theta/2\right) n/\lambda$ is the magnitude of the scattering vector for a scattering angle θ, k the Boltzmann constant, T the absolute temperature (298 K), n the refractive index of the solvent (1.332) and λ the wavelength of the laser.

3.1.4 Variational free energy calculation of the surface potential Ψ_{theo}

The net charge of a SPB can be determined through a variational free energy approach that was originally applied on polyelectrolyte stars [52,53]. This cell model can be extended to SPBs by taking into account a core of finite size [48]. Figure 3.1.2 displays the cell model on which the present calculation is based. The free energy of an isolated SPB with core radius R_c, number of chains f, a degree of polymerization N, and a brush thickness L, in correspondence to a given density, consists of the following contributions: A Hartree-type contribution describes the electrostatic potential of the SPB with a net charge Q^* [48]:

$$\frac{U_H}{kT} = \frac{Q^{*2}\lambda_B}{2e^2} u(L; R_W, R_c) \tag{3.1.1}$$

with λ_B being the Bjerrum length (0.72 nm for water at 298 K). The function $u(L; R_W, R_c)$ contains the dependency on the brush thickness L, and on the size parameters R_c and the Wigner-Seitz cell radius R_W, see Ref. [48] for more details. The next contributions account for the entropic nature of the counterions and coions. The counterions can be free or confined within the brush. Inside the brush they also can be condensed onto individual polyelectrolyte chains. In total, monovalent counterions can possess three different states, resulting in three different entropic contributions, since even the condensed ions can move along the chains [52]. The ideal entropic contributions of the counterions are obtained from

$$\frac{S_i}{kT} = \int_{V_i} \rho_i(\mathbf{r}) \left[\ln(\rho_i(\mathbf{r}) l_0^3) - 1\right] \mathrm{d}^3 r, \tag{3.1.2}$$

with $\rho_i(\mathbf{r})$ being the ion density distributions at the position $\mathbf{r}$ within the volume V_i as defined below, and l_0 being the monomer size (0.25 nm). The three different states are defined as follows: N_1 are condensed in tube volumes V_1 around each chain, N_2 are non-condensed but confined counterions within the SPB volume V_2. Condensation effects are implemented through a correlation free energy term [48,53]. The confined non-condensed counterions possess an inhomogeneous distribution inside the brush similar to the chain monomers. Simulations have shown a decay of the counterion density inside the brush that scales with the distance from the brush center as r^{-2} [53-55]. The third counterion state is defined by N_3 free counterions in the volume V_3, which is defined by the Wigner-Seitz cell radius R_W: $V_3 = (4\pi/3)[R_W^3 - (R_c + L)^3] \approx (4\pi/3)R_W^3$, since $R_c + L \ll R_W \approx 4300$nm. The details of each of the corresponding entropic terms for the monovalent cases are given in Ref. [48]. In contrast to the three possible counterion states, the coions do not enter the brush regime in our model. It has been shown that this is a reasonable approximation due to electrostatic repulsions between coions and the SPB [53]. Consequently there is only one entropic contribution from the coions obtained from Eq. [3.1.2]:

$$\frac{S_{co}}{kT} = N_{co} \left[\ln\left(\frac{N_{co}}{V_3} l_0^3\right) - 1\right] \tag{3.1.3}$$

with $N_{co} = \rho_s V_3$ being the number of coions that is defined through the number density of added salt ρ_s.

The remaining free energy contributions result from the chains [57,58]: A contribution related to the elastic energy of the chains written as

$$\frac{F_{\mathrm{el}}}{kT} = \frac{3fL^2}{2Nl_0^2}, \tag{3.1.4}$$

which stems from a Gaussian approximation of the conformational entropy of the brush chains. The self-avoidance is accounted by a Flory-type expression

$$\frac{F_{\mathrm{Fl}}}{kT} = \frac{3v(fN)^2}{8\pi\left[(R_c + L)^3 - R_c^3\right]}, \tag{3.1.5}$$

with the excluded volume parameter $v \approx l_0^3$. As usual for the case of good-solvent conditions, triplet-monomer contributions have been omitted.

The total free energy is obtained by adding up all contributions from Eqs. [3.1.1-3.1.5]. This total free energy can be minimized with respect to the number of condensed counterions N_1, the net charge Q^* and the brush thickness L. Note that in the monovalent

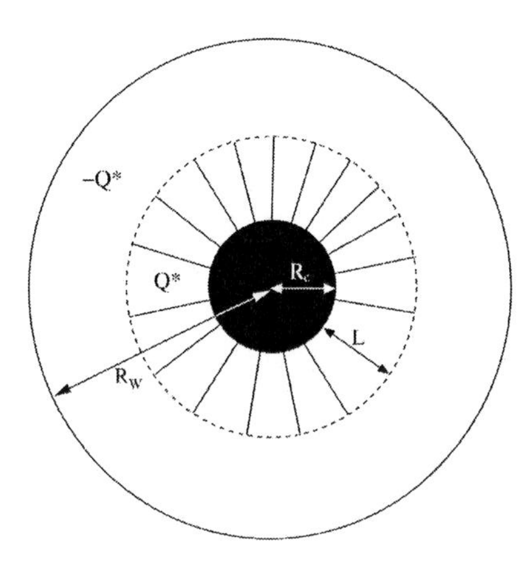

Figure 3.1.2: Scheme of the modeling of a spherical polyelectrolyte brush in its Wigner-Seitz cell. The chains are represented by lines and are grafted on the core surface (filled sphere in the center with radius R_c). The spherical particle is enclosed in a cell of radius R_W and carries the net charge Q^*. L denotes the height of the brush layer. The hydrodynamic radius R_h as measured by dynamic light scattering may be identified with $R_h = R_c + L$ (see also Ref. [43,48,56] for further details.)

case the net charge of the SPB is given by $|Q^*| = |Q_b| - e(N_1 + N_2)$, with Q_b being the bare charge of the SPB, and the neutralization condition $N_1 + N_2 + N_3 = |Q_b|/e + N_{co}$ inside the Wigner-Seitz cell. Here Q_b is given by the number of negatively charged NaSS units per SPB $N(\text{NaSS})$.

In the present study, however, trivalent salt has been added. In this case the confined monovalent counterions are exchanged by the trivalent ions [44]. Assuming a neutral SPB the exchange ratio at given ionic strength can be obtained from the corresponding Donnan-equilibrium of confined and free ions [43,44]. Here we cannot apply a simplified treatment in terms of a Donnan equilibrium since we are interested in the net charge of the SPB. Instead we have to make the following assumptions which are consistent with recent simulation studies [44]:

(i) In the case at which the number of added multivalent counterions $qN_q < |Q_b - Q^*|/e$, the confined monovalent counterions are replaced by qN_q multivalent ones (valency q). This replacement approximates the Donnan equilibrium but assumes that in this regime there are no free multivalent counterions as observed in simulations [44]. The net charge is then $|Q^*| = |Q_b| - e(qN_q + N_1 + N_2)$. The effective charge is then also given by the free monovalent ions: $|Q^*| = |Q_b| - eN_3$. The number of coions is $N_{co} = qN_q = q\rho_s V_3$.

(ii) If $qN_q \geq |Q_b - Q^*|/e$ all monovalent counterions are completely replaced by trivalent ones ($N_1 = 0$, $N_2 = 0$, $N_3 = |Q_b|/e$), and the net charge is determined through remaining trivalent ions outside the SPB: $|Q^*| = |Q_b| - e(qN_{q,o})$, with $N_{q,o}$ being the number of free multivalent ions in the volume V_3. Inside the brush there are $N_q - N_{q,o}$ multivalent ions.

In both cases we further assume that all confined multivalent ions do not contribute to the osmotic pressure, i.e. they do not possess an entropic contribution of the form Eq. [3.1.3]. This hypothesis is supported by recent simulation studies where it was demonstrated that the multivalent counterions are strongly condensed on polyelectrolyte chains, in particular when $q \geq 3$ [44,59]. Using the assumptions made in (i) and (ii) the net charge Q^* and the brush thickness L are now determined through minimization of the free energy contributions (Eqs. [3.1.1-3.1.5]) at given ionic strength $I = (1/2)q(q+1)\rho_s$. Since $q = 3$, I is identical with $I(\mathrm{EuCl_3})$ introduced in Section 3.1.5.1.

Having calculated the brush thickness L and the net charge Q^* of the SPB, the surface potential can be readily obtained. For this we treat the SPB as a sphere with radius $R_c + L$ and a surface charge density of $\Sigma = Q^*/[4\pi(R_c + L)^2]$. The asymptotic formula by Zhou is then used for the calculation of the surface potential Ψ_{theo} at given charge density Σ [60]. This approximate solution deviates less than 5% from the exact PB solution for the cases considered in this work. The theoretical prediction thus obtained can be compared to experimental results. Note that in the above case (i), the SPB with net charge Q^* can be considered as a charged particle immersed in a salt solution of an effective ionic strength $I_{eff} = (N_3 + N_{co})/(2V_3)$, consisting of monovalent ions only, since all multivalent ions are absorbed by the brush. Hence, the "effective" ionic strength accounts for the concentrations of free released counterions and coions. Beyond the transition point, $q\rho_s V_3 = qN_q \geq |Q_b - Q^*|/e$, i.e. in the case (ii), the SPB can be considered as a charged particle immersed in a mixture of non-confined multivalent ions and the relased monovalent ones, which results again in an effective ionic strength I_{eff}. The effective ionic strength enters into the screening parameter $\kappa = \sqrt{8\pi\lambda_B I_{eff}}$, which is required for the determination of the surface potential Ψ_{theo} [60].

3.1.5 Results and Discussions

The starting point of the present analysis is the assumption that the diffusional and the electrophoretic mobility refer to the same effective shear plane. Thus, R_h is the crucial quantity that determines the hydrodynamic drag of the isolated particle. The effective charge is assigned to this plane, that is, in the following the SPB are treated as a sphere with radius R_h on the surface of which the effective charge has been distributed. In general, the surface of a SPB is fuzzy and the polyelectrolyte layer exhibits marked thermal fluctuations that can be seen in small-angle X-ray scattering [61]. Hence, the surface of the SPB is defined to an uncertainty of at least 3 nm.

3.1.5.1. Hydrodynamic radius and electrophoretic mobility of SPB in solutions of $EuCl_3$

As shown in previous studies, for a given ionic strength a mixture of trivalent and monovalent salts leads to a stronger shrinking of the polyelectrolyte layer on SPB particles compared to the effect of monovalent salt [43]. Figure 3.1.3 displays the brush thickness $L = R_h - R_c$ normalized by the contour length of grafted chains L_c as a function of the total Eu^{3+} concentration in solution, into which the SPBs were immersed. The ionic strength $I(EuCl)_3$ of the solution is identical with I as introduced in Section 3.1.4. It was systematically increased from 3×10^{-6} to 1.5×10^{-4} mol/L, or in terms of κR_h, from 0.75 to 5.15. It should be noted that due to ion exchange of Na^+ ions in the brush and Eu^{3+} ions, the effective ionic strength in the equilibrium I_{eff} is always smaller than $I(EuCl)_3$. The shell shrinks for $I(EuCl)_3 > 3 \times 10^{-6}$ mol/L. For $I(EuCl)_3 = 1.5 \times 10^{-4}$ mol/L a brush thickness of 7 nm was measured which is only 10% of the thickness in salt-free water (74 nm). In all calculations of the experimental potential ζ to follow, the hydrodynamic radii given in Figure 3.1.3 are used. Cryo-TEM measurements confirmed the elongation of such brush layers in very good agreement with DLS data [42]. Thus, R_h provides a realistic measure for the extension of the surface layer.
The size of the SPBs can now be compared with results from the variational free energy calculation. The following experimentally given parameters were used for the calculations: $L_c = 82$ nm, $\sigma = 0.029$ nm^{-2} and $Q_b = 1.9 \times 10^6|e|$. The result (black line) describes the experimental data in the vicinity of the collapse . Below $I(EuCl)_3 = 10^{-5}$ mol/L (or $\kappa R_h \approx 1.25$), the measured brush thickness is larger than predicted by the theory (ca. 20%). We ascribe this difference to the polydispersity of the polymer chains in the brush layer. Given the fact that no adjustable parameter has been used in this calculation, the agreement of theory and experiment may be considered satisfactory.
If salt is added to a suspension of highly charged SPBs in water, the mobility of the particles will be balanced by two main effects: the shrinking of the shell shifts the effective shear plane nearer to the core surface and facilitates the motion of the smaller particle through the solution. On the other hand, additional salt leads to the screening of surface charges accompanied with a decreasing mobility. Moreover, the trivalent ions will neutralize a part of the charge as discussed above. For the SPBs, both effects are superimposed. Figure 3.1.3 also shows the electrophoretic mobility μ of the brush particles (filled circles). During the collapse of the brush layer up to $I(EuCl)_3 < 2 \times 10^{-5}$ mol/L, μ changes from -4.1 to -3.5 μm $\times$ cm $\times$ V^{-1}s^{-1}. Further increase of $I(EuCl_3)$ does not significantly alter the brush thickness but it does change the electrophoretic mobility. Hence, this effect must be attributed to a partial neutralization of the brush layer by the trivalent ions. This conclusion is based on calculations that show that the collapse transition takes place at an ionic strength (1.5×10^{-5} mol/L) where most of the Na^+ counterions have been replaced by trivalent europium ions .

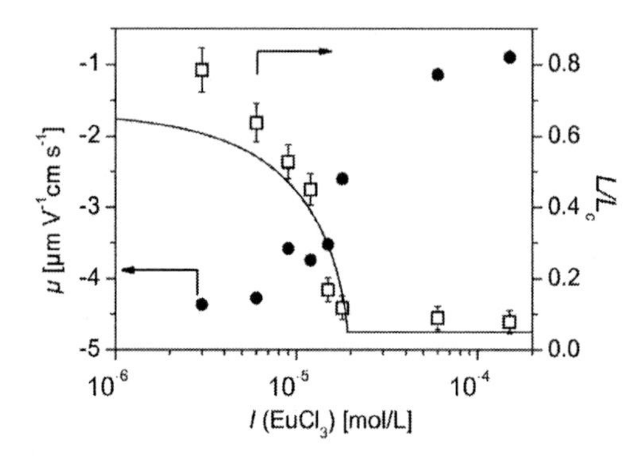

Figure 3.1.3: Variation of experimental electrophoretic mobility μ (filled circles, left ordinate) and reduced brush thickness $(R_h - R_c)/L_c$ (open squares, right ordinate) of SPBs with ionic strength (total amount of $EuCl_3$). The black line is the result from variational free energy calculations. $I(EuCl)_3 = 3 \times 10^{-6}$ mol/L corresponds to $\kappa R_h = 0.75$, and 1.5×10^{-4} mol/L to $\kappa R_h = 5.15$. κ reflects the effective ionic strength I_{eff} in the equilibrium.

3.1.5.2. Experimental ζ-potential of SPB in solutions of $EuCl_3$

Since an identical effective shear plane is assumed for both the hydrodynamic radius R_h and for the electrophoretic mobility μ, the ζ-potential of the SPBs can be determined from these experimental values. For this purpose the theory of O'Brien and White [9] was used. For calculations the MPEK-0.01 software package developed by Hill (compare Ref. [62]) was applied for the case of solid charged and uncoated spheres similar to that described by O'Brien and White [19]. For the calculations we employed experimental parameter values, i.e. $T = 298$ K and $\eta = 0.891$ cP (viscosity of water). For the molar (equivalent) conductivities of Na^+, Cl^-, and $1/3Eu^{3+}$, we used 50.08×10^{-4} m^2S mol^{-1}, 76.31×10^{-4} m^2S mol^{-1}, and 67.8×10^{-4} m^2S mol^{-1}, respectively [63].
In addition, the equilibrium ionic strength I_{eff} or the inverse Debye length κ is needed. The approach to calculate the concentration of the different ions in equilibrium is as follows (see above): i) Ion exchange of Eu^{3+} (from the bulk into the brush) and Na^+ counterions (from the brush into the bulk solution) takes place until every Na^+ counterion is replaced by the equivalent amount of Eu^{3+} ions. ii) The amount of Na^+ counterions is given by the SPB concentration and the number of NaSS-units per SPB. iii) Free Eu^{3+} ions are present in the bulk phase if all Na^+ ions have been replaced in the brush layer. The experimental ζ-potential was determined from the corresponding mobility by interpolation using calculated pairs of ζ/μ-values for $-150 \leq \zeta[\text{mV}] \leq 0$. For $\kappa R_h > 3$, theory predicts a pronounced maximum of $\mu(\zeta)$ [9,11]. As a result of this there may be two possible values for the potential if ζ is sufficiently high. Howe-

ver, this ambiguity is not relevant here because of the low experimental μ-values. In Figure 3.1.4 the ζ-potential and the electrophoretic mobility are presented in reduced units: $y = \zeta e/(kT)$ and $E = 3e\eta\mu/(2kT\epsilon)$ (ϵ solvent permittivity). In these experiments κR_h varies between 0.75 and 5.15, respectively. For the lowest ionic strength, or $\kappa R_h = 0.75$, the experimental potential $y = -4.25$ ($\zeta = -110$ mV) predicts a high electrostatic stability of the particles. In this regime, no free Eu^{3+} ions are present in the bulk phase. After the collapse transition at $\kappa R_h \approx 1.3$ ($I(EuCl)_3 > 1.5 \times 10^{-5}$ mol/L) free Eu^{3+} ions are also present in the solution. The ζ-potential is finally reduced to -15 mV or $y = -0.63$ ($I(EuCl)_3 = 10^{-4}$ mol/L), which corresponds to $e\zeta$ smaller than kT. This low potential agrees with the observation that trivalent ions destabilize the SPB [46]. Note that the trend of E is the same as for y. Both quantities decrease rapidly when the collapse transition takes place.

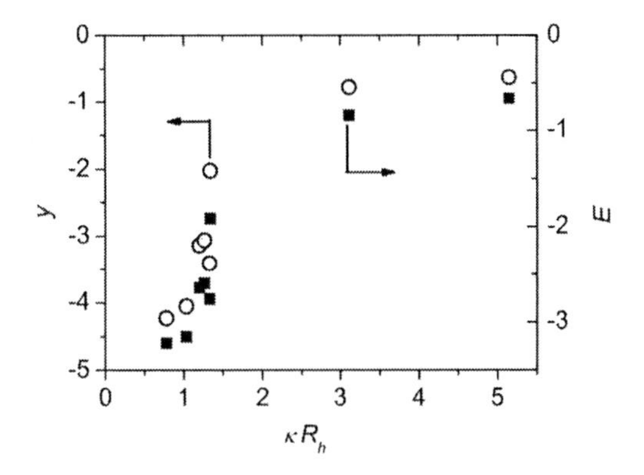

Figure 3.1.4: Variation of reduced experimental ζ-potential $y = \zeta e/(kT)$ (open circles, left ordinate) and reduced electrophoretic mobility $E = 3e\eta\mu/(2kT\epsilon)$ (filled squares, right ordinate) of the SPBs as a function of κR_h. κ reflects the effective ionic strength I_{eff} in the equilibrium.

In the following the limits of small and large κR_h will be discussed. In all cases, the brush layer is smaller than the electric double layer. When only Eu^{3+} counterions are present in the brush layer (valid for $\kappa R_h > 1.4$), no electroosmotic flow can develop, since trivalent ions have been shown to be osmotically inactive [44]. The Brinkman screening length l_B varies between 0.78 and 0.62 nm thus pointing to an almost impermeable layer ($1.4 \leq \kappa R_h \leq 5.15$). For the calculation of l_B, an average polymer segment density (n_S) of 1.34 or 2.07 mol/L and $a_S = 0.095$ nm [20] for the size of the Stokes resistance centers was used. F_S was determined using Eq. (10) in Ref. [20] valid for volume fractions $\phi_S = 4/3\pi a_S^3 n_S < 0.4$. For $\kappa R_h < 1.4$ the influence of an electroosmotic flow cannot be ruled out, since 20 to 80 % of Na^+ ions are in the brush layer with $l_B = 0.96$ nm ($n_S = 0.91$ mol/L) or $l_B = 2.59$ nm ($n_S = 0.13$ mol/L). However, the mobility of theNa^+

ions inside a dense brush layer is strongly diminished due to counterion condensation which was also found in Ref. [34]. It is noteworthy that the largest l_B value (2.6 nm) is the lower limit of accuracy of the DLS measurement (ca. 1.5% of particle size). As l_B is sufficiently low, the fluid flow inside the brush layer can be neglected and the SPBs can be treated as impenetrable charged spheres.

3.1.5.3. Comparison of Ψ_{theo} and ζ for the SPB in the presence of $EuCl_3$

In the following we present the comparison of the surface potential Ψ_{theo} obtained from variational free energy calculations with the experimental ζ-potential. Ψ_{theo} was calculated from the surface charge density $\Sigma = Q^*/(4\pi R_h^2)$ of the SPBs obtained from the analytic treatment described in Section 3.1.4.
The effective charge Q^* of each SPB in aqueous solution is shown in Figure 3.1.5 as a function of the dimensionless particle size κR_h. For very low concentrations of $EuCl_3$, the SPBs exhibit the highest effective charge ($Q^* \approx 3800e$). Note that with a decreasing particle charge, κR_h first increases, but decreases near its value of 1.3. This is due to the rapid shrinking of the brush layer (and thus R_h) at the corresponding ionic strength although κ increases moderately. A discharging of the brush layer due to trivalent ions far above the collapse transition at $\kappa R_h = 5.15$ or $I(\text{EuCl})_3 = 1.5 \times 10^{-4}$ mol/L is monitored by a decrease of Q^* by a factor of five. This effective charge ($600e$) corresponds to about 0.04% of the bare charge Q_b [46].

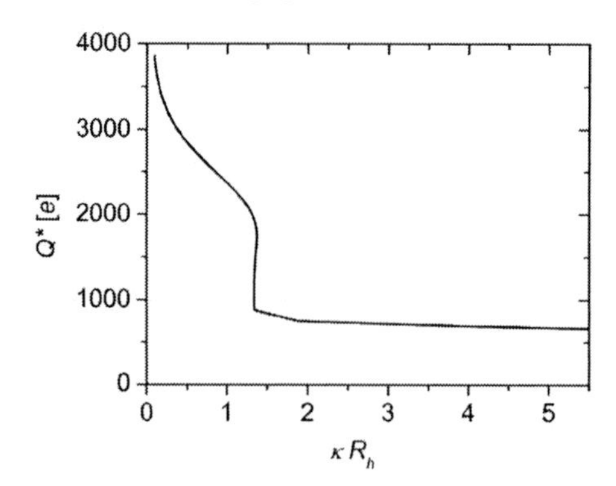

Figure 3.1.5: Effective charge of the SPBs (Q^*) from variational free energy calculations in units of the elementary charge e as a function of the reduced particle size κR_h. κ reflects the effective ionic strength I_{eff} in the equilibrium.

Figure 3.1.6 displays a comparison of both Ψ_{theo} and ζ. As is seen from this comparison, Ψ_{theo} (solid line) and ζ (hollow circles) agree within experimental error for $\kappa R_h > 0.78$, or $I(EuCl_3) > 3 \times 10^{-6}$ mol/L. Note that no adjustbale parameters were used in the variational free energy calculations. Thus, the underlying assumptions of the simplified treatment presented here seem to be fully corroborated by experimental data. An important point which should be kept in mind is the strong correlation of the trivalent counterions Eu^{3+} to the polyelectrolyte chains. This complexation in return is responsible for strong decrease of the thickness L of the surface layer [44,59]. The present comparison of theory with data related to the electrophoretic mobility furthermore corroborates the assumption that this strong correlation suppresses all electroosmotic effects within the surface layer.

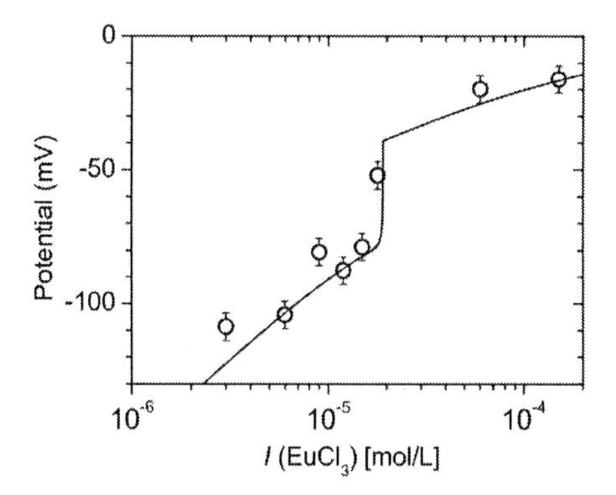

Figure 3.1.6: Comparison of the surface potential Ψ_{theo} for the SPBs in $EuCl_3$ solutions as determined by variational free energy calculation (solid line) with the ζ-potential derived from the experimental electrophoretic mobilities (open circles). When the brush layer is fully collapsed on the core particle, Ψ_{theo} and ζ are reduced to ca. -15 mV. $I(\text{EuCl})_3 = 3 \times 10^{-6}$ mol/L corresponds to $\kappa R_h = 0.75$, and $I(\text{EuCl})_3 = 1.5 \times 10^{-4}$ mol/L to $\kappa R_h = 5.15$. κ reflects the effective ionic strength I_{eff} in the equilibrium.

3.1.6 Conclusions

In conclusion, we show that the experimental ζ-potential of spherical polyelectrolyte brushes in the presence of trivalent salt is in quantitative agreement with results from variational free energy calculations Ψ_{theo} without adjustable parameters. The collapse transition in the brush layer is due to a partial neutralization of the charges by the trivalent ions leading to a steep decrease of the absolute value of Ψ_{theo} and ζ. Thus, there is strong evidence that the SPBs behave as compact colloids rather than soft ones in the presence of trivalent counterions. This finding is attributed to the suppression of electroosmotic effects within the polyelectrolyte layer by the strong correlation of the trivalent ions to the polyelectrolyte macroion.

Acknowledgments

This work was financially supported by the Deutsche Forschungsgemeinschaft. M. H. and C. S. gratefully acknowledge the Bavarian Elite Network (ENB) Study Program "Macromolecular Science". M. H. thanks Prof. Reghan J. Hill for kindly providing the MPEK-0.01 software package.

Appendix. List of Symbols

1. a_S (m): Stokes radius of a polymer segment
2. D_T (m^2s^{-1}): translational diffusion coefficient
3. e (C): elementary charge
4. E: reduced electrophoretic mobility
5. F_{el} (J): free energy contribution of the elastic chains
6. F_{FL} (J): free energy contribution of the self-avoidance
7. F_S: dimensionless drag coefficient
8. I (m^{-3}): ionic strength of $EuCl_3$ before addition of the SPBs; $I = I(EuCl_3)$
9. $I(EuCl_3)$ (mol L^{-1}): see I
10. I_{eff}: effective ionic strength outside the SPB in the equilibrium
11. K ($kgm^{-3}s^{-1}$): friction coefficient of polymer segments
12. L (m): length of grafted chains in solution
13. l_0 (m): monomer size
14. l_B (m): Brinkman screening length
15. L_c (m): contour length
16. n: refractive index
17. n_S (mol/L): density of polymer segments
18. N: number of monomer segments per chain
19. N(NaSS): number of NaSS units per SPB
20. n_S (mol/L): density of polymer segments

21. N_i: number of counterions in state i ($i = 1$: condensed in V_1, $i = 2$: confined in V_2, $i = 3$ free in V_3)
22. N_q: number of added multivalent counterions (total)
23. $N_{q,0}$: number of multivalent counterions outside the brush
24. q: valence of multivalent counterion ($q = 3$)
25. $|\vec{q}|$ (m^{-1}): magnitude of the scattering vector
26. Q_b (e): bare surface charge
27. Q^* (e): effective surface charge
28. R_c (m): radius of the core particle
29. R_W (m): radius of the Wigner-Seitz cell
30. R_h (m): hydrodynamic radius ($R_c + L$)
31. S_{co} (J): ideal entropic contribution of the coins
32. S_i (J): ideal entropic contribution of counterion species i to U_H ($i = 1$ condensed, $i = 2$ confinded, $i = 3$ free)
33. U_H (J): electrostatic potential of the SPB
34. v (m^3): excluded volume parameter
35. V_i (m^{-3}): location for counterion species i ($i = 1$ tube volume around a polymer chain, $i = 2$ volume of the brush layer, $i = 3$ volume of the Wigner-Seitz cell)
36. y: reduced ζ-potential
37. Γ (s^{-1}): relaxation frequency
38. ϵ ($\mathrm{J}^{-1}\mathrm{C}^2\mathrm{m}^{-1}$): solvent permittivity
39. Φ_S: volume fraction
40. η (Pa s): solvent viscosity
41. η_W (Pa s): dynamic viscosity
42. κ (m^{-1}): inverse Debye length at I_{eff}
43. λ (m): wavelength of the laser light

44. λ_0^{-1} (m): penetration length
45. λ_B (m): Bjerrum length
46. L (m): length of grafted chains in solution
47. μ ($\mathrm{m^2V^{-1}s^{-1}}$): electrophoretic mobility
48. θ (^{O}): scattering angle
49. $\rho_i(\mathbf{r})$ ($\mathrm{m^{-3}}$): ion density distribution at the position $\mathbf{r}$ in a volume V_i
50. ρ_S ($\mathrm{m^{-3}}$): number density of added salt
51. σ ($\mathrm{m^{-2}}$): grafting density of attached polymer chains
52. Σ ($e\mathrm{m^{-2}}$): surface charge density
53. Ψ_{theo} (V): surface potential from variational free energy calculations
54. ζ (V): electrokinetic or $\zeta-$ potential

3.1.7 References and Notes

[1] W. B. Russel, D. A. Saville, W. R. Schowalter, Colloidal Dispersions, Cambridge University Press, Cambridge, 1999 (1. Ed.).

[2] R. J. Hunter, Zeta potential in colloid science - Principles and Applications, Academic Press, London, 1981 (1. Ed.).

[3] P. H. Wiersema, A. L. Loeb, J. Th. G. Overbeek, J. Colloid Interf. Sci. 22 (1966) 78-99.

[4] R. J. Hunter, Foundations of Colloid Science, Oxford University Press, New York, 2004 (2. Ed.), 373-433.

[5] A. V. Delgado, F. Gonzalez-Caballero, R. J. Hunter, L. K. Koopal, J. Lyklema, J. Colloid Interface Sci. 309 (2007) 194-224.

[6] E. Hückel, Phys. Z. 25 (1924) 204-210.

[7] M. von Smoluchowski, Z. Phys. Chem. 92 (1918) 129-168.

[8] D. C. Henry, Proc. R. Soc. Lond. A. 133 (1931) 106-129.

[9] R. W. O'Brien, L. R. White, J. Chem. Soc. Faraday Trans. II 74 (1978) 1607-1627.

[10] A. S. Russel, P. J. Scales, C. S. Mangelsdorf, S. M. Underwood, Langmuir 11 (1995), 1112-1115.

[11] M. R. Gittings, D. A. Saville, Coll. Surf. A: Physicochem. Eng. Asp. 141 (1998) 111-117.

[12] S. S. Dukhin, R. Zimmermann, C. Werner, Coll. Surf. A: Physicochem. Eng. Asp. 195 (2001) 103-112.

[13] H. Ohshima, T. W. Healy, L. R. White, J. Chem. Soc. Faraday Trans. II 79 (1983) 1613-1628.

[14] H. Ohshima, Coll. Surf. A: Physicochem. Eng. Asp. 267 (2005) 50-55.

[15] J. F. L. Duval, H. Ohshima, Langmuir 22 (2006) 3533 - 3546.

[16] S. H. Behrens, D. I. Christl, R. Emmerzael, P. Schurtenberger, M. Borkovec, Langmuir 16 (2000) 2566-2575.

[17] M. Kobayashi, Colloid Polym. Sci. 286 (2008), 935-940.

[18] M. Quesada-Pérez, E. González-Tovar, A. Martín-Molina, M. Lozada-Cassou, R. Hidalgo-Álvarez, Coll. Surf. A: Physicochem. Eng. Asp. 267 (2005) 24-30.

[19] R. J. Hill, D. A. Saville, W. B. Russel, J. Colloid Interf. Sci 258 (2003) 56-74.

[20] R. J. Hill, D. A. Saville, Colloids Surf. A 267 (2005) 31-49.

[21] P. Borget, F. Lafuma, C. Bonnet-Gonnet, J. Colloid Interf. Sci. 285 (2005) 136-145.

[22] M. Ballauff, Progr. Polym. Sci. 32 (2007) 1135-1151.

[23] J. J. Lietor-Santos, A. Fernandez-Nieves, Adv. Colloid Interf. Sci. 147-148 (2009) 178-185.

[24] Y. Lu, M. Ballauff, Polymer 48 (2007) 1815-1823.

[25] H. Ohshima, K. Makino, T. Kato, K. Fujimoto, T. Kondo, H. Kawaguchi, J. Colloid Interf. Sci. 159 (1993), 512-514.

[26] D.A. Saville, J. Colloid Interface Sci. 222 (2000) 137-145.

[27] H. Ohshima, Adv. Colloid Interface Sci. 62 (1995) 189-235.

[28] D. L. Koch, A. S. Sangani, J. Fluid Mech. 400 (1999) 229-263.

[29] S. S. Dukhin, R. Zimmermann, C. Werner, J. Colloid Interf. Sci. 286 (2005) 761-773.

[30] S. S. Dukhin, R. Zimmermann, C. Werner, J. Colloid Interf. Sci. 274 (2004) 309-318.

[31] S. S. Dukhin, R. Zimmermann, C. Werner, J. Colloid Interf. Sci. 328 (2008) 217-226.

[32] S. S. Dukhin, R. Zimmermann, C. Werner, J. Phys. Chem. B. 111 (2007) 979-981.

[33] J. F. L. Duval, H. P. van Leeuwen, Langmuir 20 (2004), 10324-10336.

[34] R. Zimmermann, W. Norde, M. A. Cohen Stuart, C. Werner, Langmuir 21 (2005) 5108-5114.

[35] M. Ballauff, O. Borisov, Curr. Opinion Colloid Interf. Sci. 11 (2006) 316-323.

[36] N. Volk, D. Vollmer, M. Schmidt, W. Oppermann, K. Huber, Adv. Polym. Sci. 166 (2004) 29-65.

[37] R. C. Advincula, W. Brittain, K. Caster, J. Rhein, Polymer Brushes, Wiley-VCH, Weinheim, 2004.

[38] P. Pincus, Macromolecules 24 (1991) 2912-2919.

[39] O. V. Borisov, T. M. Birshtein, E. B. Zhulina, Journal de Physique II 1 (1991) 521-526.

[40] B. Das, X. Guo, M. Ballauff, Progr. Colloid Interf. Sci. 121 (2002) 34-38.

[41] H. Ahrens, S. Forster, C. A. Helm, Phys. Rev. Lett. 81 (1998) 4172-4175.

[42] A. Wittemann, M. Drechsler, Y. Talmon, M. Ballauff, J. Am. Chem. Soc. 127 (2005) 9688-9689.

[43] Y. Mei, K. Lauterbach, M. Hoffmann, O. V. Borisov, M. Ballauff, A. Jusufi, Phys. Rev. Lett. 97 (2006) 158301.

[44] Y. Mei, M. Hoffmann, M. Ballauff, A. Jusufi, Phys. Rev. E. 77 (2008) 031805.

[45] K. J. Sasaki, S. L. Burnett, S. D. Christian, E. E. Tucker, J. F. Scamehorn, Langmuir 5 (1989) 363-369.

[46] C. Schneider, A. Jusufi, R. Farina, F. Li, P. Pincus, M. Tirrell, M. Ballauff, Langmuir 24 (2008) 10612-10615.

[47] Eu^{3+} ions have been used instead of La^{3+} ions as they can be used as fluorescent probes, which is of interest for further experiments, see R. P. Fisher and J. D. Winefordner, Anal. Chem. 43 (1971) 454-455.

[48] A. Jusufi, C. N. Likos, M. Ballauff, Colloid Polym. Sci. 282 (2004) 910-917.

[49] J. H. Kim, M. Chainey, M. S. El-Aasser, J. W. Vanderhoff, J. Polym. Sci. A 30 (1992) 171-183.

[50] X. Guo, A. Weiss, M. Ballauff, Macromolecules 32 (1999) 6043-6046.

[51] M. Schrinner, B. Haupt, A. Wittemann, Chemical Engineering Journal 144 (2008), 138-145.

[52] A. Jusufi, C. N. Likos, H. Löwen, Phys. Rev. Lett. 88 (2002) 018301.

[53] A. Jusufi, C. N. Likos, H. Löwen, J. Chem. Phys. 116 (2002) 11011-11027.

[54] H. Wang, A. R. Denton, Phys. Rev. E 70 (2004) 041404.

[55] A. Jusufi, J. Chem. Phys. 124 (2006) 044908.

[56] X. Guo, M. Ballauff, Phys. Rev. E. 64 (2001) 051406.

[57] S. Alexander, Journal de Phys. 38 (1977) 977-981.

[58] P. G. de Gennes, Macromolecules 13 (1980) 1069-1075.

[59] R. Ni, D. Cao, W. Wang, A. Jusufi, Macromolecules 41 (2008) 5477-5484.

[60] S. Zhou, Colloid Interf. Sci. 208 (1998), 347-350.

[61] N. Dingenouts, M. Patel, S. Rosenfeldt, D. Pontoni, T. Narayanan, M. Ballauff, Macromolecules 37 (2004) 8152-8159.

[62] The manual for the MPEK-software package can be download from Reghan J. Hill.<http://people.mcgill.ca/files/reghan.hill/MPEK-0.02.pdf>.

[63] Lide, D. R. (Ed.), CRC Handbook of Chemistry and Physics, 76th ed., CRC Press: Boca Raton, Fl, 1995; Chapter 5, p 90.

3.2 Dumbbell-Shaped Polyelectrolyte Brushes Studied by Depolarized Dynamic Light Scattering

Martin Hoffmann, Yan Lu, Marc Schrinner and Matthias Ballauff

Physikalische Chemie I, University of Bayreuth, 95440 Bayreuth, Germany

Ludger Harnau

Max-Planck-Institut für Metallforschung, Heisenbergstraße 3, D-70569 Stuttgart (Germany) and Institut für Theoretische und Angewandte Physik, Universität Stuttgart, Pfaffenwaldring 57, D-70569 Stuttgart (Germany)

*Corresponding author: Matthias.ballauff@uni-bayreuth.de

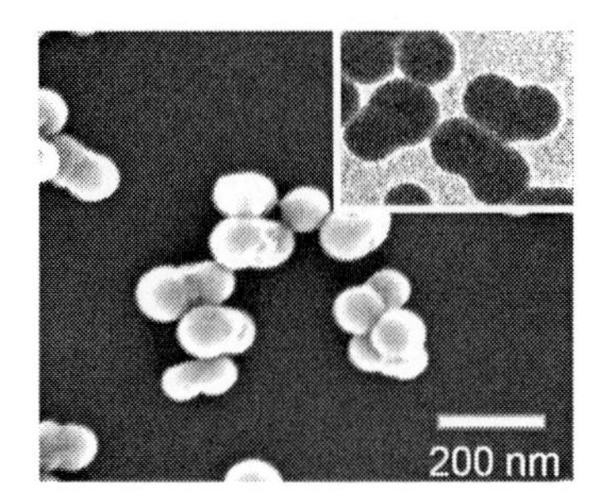

Published in the *Journal of Physical Chemistry B* **2008**, *112*, 14843-14850.

3.2.1 Abstract

We present the synthesis and comprehensive characterization of dumbbell-shaped polyelectrolyte brushes (DPB). The core of these particles consists of poly(methyl methacrylate) (PMMA) and poly(styrene) onto which a dense brush shell of poly(styrene sulfonate) is grafted. The morphology of DPB particles is studied in solution by cryogenic-transmission electron microscopy. We demonstrate that well-defined DPB are generated that react to external stimuli such as surfactant and salt concentration. The rotational diffusion and collective relaxations of the DPB particles were monitored by depolarized dynamic light scattering (DDLS). Here we found a new relaxation mode in the DDLS-signal that can be ascribed to collective fluctuations of the polyelectrolyte layer affixed to the surface of the dumbbells.

3.2.2 Introduction

If polyelectrolyte chains are densely grafted to colloidal spheres, spherical polyelectrolyte brushes (SPB) result.[1-4] Immersed in water, the polyelectrolyte layer attached to the surface of the core particles will swell if the ionic strength in the system is low. This remarkable effect[5] is due to the high osmotic pressure of the counterions which are confined within the brush layer. If, on the other hand, the ionic strength is high, the polyelectrolyte layer will collapse. Since most of the properties of colloidal particles are determined by their surface, such spherical polyelectrolyte brushes present highly versatile systems. Applications suggested so far include the use of these particles as carrier for catalytically active metal nanoparticles,[6,7] enzymes,[8,9] and proteins.[10,11]
In this paper, we report the synthesis of anisotropic colloidal brush particles. Anisotropic colloidal particles have recently attracted great interest due to their additional orientational degrees of freedom.[12,13] Hence, rod-[14] or disklike[15,16] structures, nonspherical architectures,[17,19] and particles with ellipsoidal geometry[20] have been synthesized and characterized. Dumbbell-shaped particles consisting of two spherical units affixed to each other have been of particular interest.[21,22] Seeded emulsion polymerization has been used in order to obtain colloidal dumbbells.[23-25] Often, the resulting particles are rather large (500 nm to 10 μm diameter). In addition to this, to the authors' best knowledge, no attempt has been made to modify the surface of such anisotropic colloids by grafted polymers.
Here we present the synthesis and a comprehensive characterization of dumbbell-shaped polyelectrolyte brush (DPB) particles. The core of these particles (200 nm diameter) consists of poly(methyl methacrylate) (PMMA) and poly(styrene) (PS) from which a dense brush shell of poly(styrene sulfonate) is grafted (Scheme 3.2.1). We hence extend previous work[3,4] on spherical SPB to nonspherical DPB. As in the case of the SPB the synthesis of the surface layer proceeds through a grafting-from technique (Scheme 3.2.1): A thin layer of photoinitiator is affixed on the core particles to generate radicals under UV radiation, which can initiate the photoemulsion polymerization of the functional monomer sodium styrenesulfonate (NaSS). The morphology of DPB particles is studied in solution, that is, in situ, by cryogenic-transmission electron microscopy (cryo- TEM). In order to investigate the diffusion of the geometric anisotropic particles in solution at various salt concentrations, we use depolarized dynamic light scattering (DDLS). This technique simultaneously probes the translational and the rotational diffusion coefficient for geometric or optic anisotropic particles.[26] So far, DDLS has been applied to prolate latex particles,[27,28] anisotropic copolymer spheres,[29-31] oligonucleotides,[32,33] DNA fragments,[34] the tobacco mosaic virus,[35] fibers,[36,37] and other systems.[38-40] Here we demonstrate that DDLS is suitable to probe the rotational motion of the dumbbell-shaped polyelectrolyte brushes with great precision. Moreover, DDLS gives additional information about the fluctuations of the surface layer of the attached polyelectrolyte chains.

Scheme 3.2.1: Scheme for the Preparation of Dumbbell-Shaped Polyelectrolyte Brushes (DPB)[a]

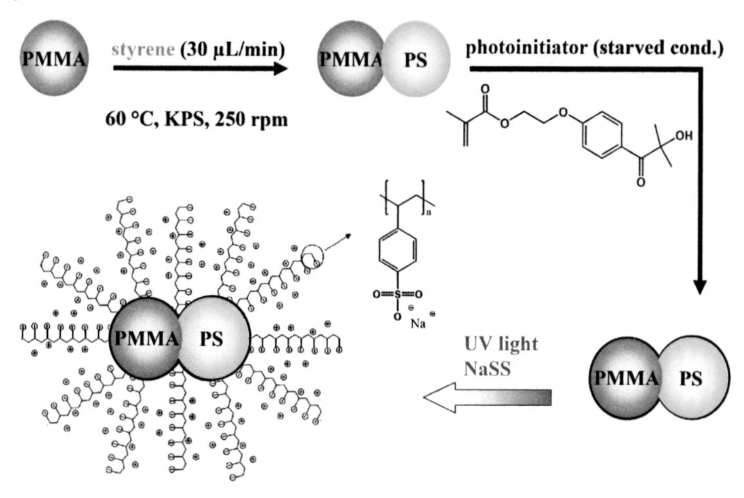

[a] Poly(methyl methacrylat) (PMMA) particles are first prepared by conventional emulsion polymerization. The dumbbell morphology is formed when adding styrene under starved conditions (30 μL/min). In the next step, these core particles are covered with a thin layer of photoinitiator HMEM. In the last step, the shell of the polyelectrolyte brushes is formed by photoemulsion polymerization; UV irradiation of the core particles generates radicals at their surface with initiate the radical polymerization of sodium styrene sulfonate (NaSS).

3.2.3 Experimental Section

3.2.3.1. Materials

PMMA-seeds were prepared by conventional emulsion polymerization (see Scheme 3.2.1). Sodium dodecylbenzene sulfonate (1.499 g; Aldrich, > 80%) was dissolved in Millipore water (353 g) under stirring. Then methyl methacrylate (100.05 g; Aldrich, > 99%, destabilized) was added under nitrogen. The polymerization was started by adding potassimum peroxodisulfate (0.401 g; KPS, Merck, p. A.) dissolved in water (20 g). The mixture was stirred at 70 °C for 15 h at 280 rpm. After purification by serum replacement (5 days), the PMMA latex (5.38 g PMMA particles in 147 g of water) was heated to 60 °C under nitrogen, and then the second-stage polymerization was carried out by dropping styrene (13.51 g; Aldrich, > 99%, destabilized over Al_2O_3 column) and initiator solution (0.3 g KPS in 15 g of water) to the seed particles simultaneously (35 μL/min). After the last drop, the temperature was raised to 70 °C and maintained for 2 h. Purification was carried out by ultrafiltration (5 days) and a uniform particle fraction was obtained with an ultracentrifuge (Allegra 64R, Beckman Coulter). To cover the core particles with a layer of photoinitiator, 2-[*p*- (2-hydroxy-2-methylpropiophenone)]-ethyleneglycol methacrylat (HMEM), was synthesized as described previously.[1,2] The purified core particles (2.63 g in 32.9 g of water) were heated at 70 °C in the absence of oxygen under stirring. The initiator (0.05 g KPS in 2.2 g of water) was added 10 min before the photoinitiator solution (0.1 g HMEM in 2.2 mL of acetone), both at a speed of 50 μL/min. After an additional 30 min reaction and cooling down to room temperature, the latex was purified by serum replacement against the 10-fold volume of water. The photoemulsion polymerization was carried out with a UV lamp (Heraeus TQ 150 Z3, range of wavelength 200-600 nm) as reported previously.[1,2] The dumbbell-shaped core particles (1.05 g in 66.5 g of water) were mixed with sodium styrenesulfonate (Aldrich, p. a.; 0.71 g in 16.1 g of water) under stirring. After one hour of irradiation, the brush latex was purified by ultrafiltration (20-fold amount of water). For the spherical polyelectrolyte brushes (SPB) the core particles were synthesized in an emulsifier free emulsion polymerization according to ref.[41] using Millipore water (2625 g), NaSS (0.81 g), styrene (375 g), $NaHCO_3$ (2.02 g; Merck, p. A.), Na_2SO_3 (1.51 g; Merck, p. A.) and KPS (3.24 g) under constant stirring (260 rpm). After 90 min. of reaction at 80 °C, the mixture was cooled down to 70 °C and 33.2 g of a concentrated solution of HMEM in acetone (79 wt%) was added within 1.5 h under starved conditions. After 4 more hours under stirring, the latex was purified by ultrafiltration against Millipore water extensively. The photoemulsion polymerization was carried out as given in ref.[1] with purified latex coated with photoinitiator (704 g in water; 7.11 wt%), NaSS (44.1 g) and Milliporewater (1297 g).

3.2.3.2. Methods

Cryogenic-transmission electron microscopy (cryo- TEM) specimens were prepared by vitrification of thin liquid films (0.2 wt% in 0.1 mM *N*-dodecyl pyridinium chloride, Merck, > 99 wt%) supported on a TEM copper grid (Agar G 600HH Cu, Polyscience) in liquid ethane at its freezing point. Examinations were carried out at a Zeiss EM 922Omega EFTEM (Zeiss NTS GmbH, Oberkochen, Germany) at a temperature around 90 K operating at 200 kV. The image was recorded digitally by a bottom-mounted charge-coupled device (CCD) camera system (UltraScan 1000, Gatan) and processed with a digital imaging processing system (Digital Micrograph 3.10 for GMS 1.5, Gatan). Elastic brightfield TEM was performed on the same electron microscope. Field emission scanning electron microscopy (FESEM) was done with a LEO 1530 Gemini microscope equipped with a field emission cathode (acceleration voltage 2000 V). The diluted sample was dried on a Si-wafer and sputtered with Pt (Cressington Sputter Coater 208 HR). Dynamic light scattering experiments were done with an ALV/DLS/SLS-5000 compact goniometer system (Peters) equipped with a He-Ne laser (632.8 nm) and a thermostat

(Rotilabo, ±0.1 °C) at 25 °C. The samples were diluted with water (η = 0.891 cp)[42] or aqueous sodium chloride solutions (1 to 100 mM, for viscosities see ref.[43]). The concentration of the core particles was 0.1 wt% in water and 0.01 wt% in 10 mM NaCl and of the DPB particles 0.005 wt% for all salt concentrations. The samples were filtered through 1 μm nylon filters in dust free glass cuvettes. Those were immersed in an index matching *cis*-decaline bath. In contrast to toluene, *cis*-decaline does not change the polarization plane of the laser light. For DDLS we measured three runs of 900 s for scattering angles between 20 and 80° with angular steps of 2, 5, or 10°, respectively. The scattered light passed through a Glan Thomson polarizer with an extinction ratio better than 10^{-5}. The relaxation frequencies were obtained by CONTIN-analysis of the intensity autocorrelation functions.

3.2.4 Results and Discussions

3.2.4.1. Dumbbell-Shaped Core Particles

In the first part of this section, we discuss the synthesis of the dumbbell-shaped core particles and their analysis by means of electron microscopy (FESEM) and DDLS. The core particles are made in an emulsifier-free seeded emulsion polymerization of PS onto PMMA-particles. It is known that the morphology of such particles depends on thermodynamic[44,45] and kinetic parameters. [46] We found that a relatively high interfacial tension between the surface of the PMMA seed particles and the styrene monomer favors the dumbbell morphology. For this purpose, we used PMMA seed particles, which were obtained in a conventional emulsion polymerization and carefully dialyzed against pure water to remove the free surfactant. The proper choice of surfactant is crucial to avoid spherical core shell morphologies when styrene is slowly added to the seed particles. Thus, the dumbbell morphology was only obtained with sodium dodecyl styrene sulfonate (SDBS). In order to achieve a kinetic control of the reaction, styrene was added under starved conditions (35μL/min) at a low reaction temperature (60 °C). This reduces the diffusion of styrene into the seed particle and helps to achieve a well-defined dumbbell-like shape. In addition to the 200 nm sized dumbbell-shaped core particles, we also found a small amount of spherical polymer particles (30 nm). We attribute this to the presence of surfactant that could not be removed by dialysis of the PMMA cores prior to the addition of styrene. These smaller particles could be removed from the dumbbell-shaped cores during ultrafiltration and ultracentrifugation.
Figure 3.2.1 displays a FESEM micrograph of PMMA/PS dumbbell core particles with a diameter of approximately 200 nm. The particles exhibit a uniform smooth surface. It is noteworthy that no radiation damage could be observed during the investigations with TEM (inset of Figure 3.2.1), which would have to be expected for bare PMMA-cores.[47,48] This leads to the conclusion of a poly(styrene)-rich particle surface, which will be important to give a uniform layer of HMEM on the core particles.

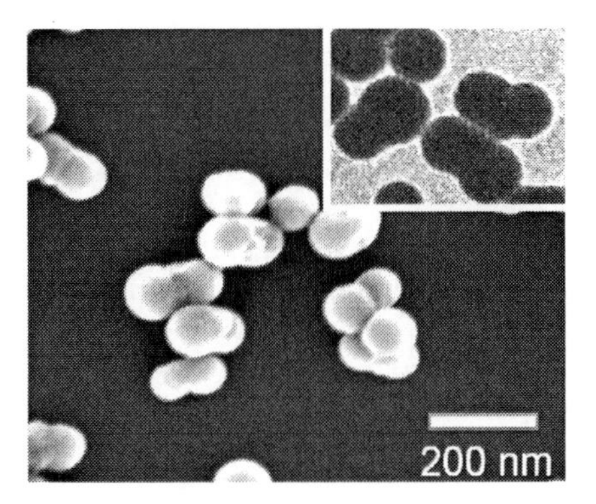

Figure 3.2.1: FESEM and TEM micrograph (inset, same scale bar) of PMMA/PS dumbbell particles.

In the following, we discuss the results of the translational and rotational motion of the dumbbells by depolarized dynamic light scattering (DDLS).[26] DDLS allows us to measure both the translational (D^T) and the rotational diffusion coefficient (D^R) of anisotropic particles. The scattered intensity autocorrelation function is a sum of two (or more) discrete exponential decays, where the slow relaxation mode may characterize translational diffusion while the faster relaxation modes may be related to rotational diffusion or internal motions of the particle. However, for a given wavelength λ and a solvent with refractive index n, the contribution of the fast modes to the intensity autocorrelation function of a conventional dynamic light scattering (DLS) experiment is often considerably smaller than the contribution of the slow mode for scattering vectors q which are comparable to the dimension of the anisotropic particle ($q = 4\pi n \sin(\theta/2)/\lambda$). Hence, in most cases DLS is not suitable to investigate the rotation of anisotropic particles. Therefore one must study the fluctuations of the depolarized light scattering intensity in order to gain insight into the rotational dynamics of the particles.
If both the diffusion coefficients D^T and D^R are known, the size parameters of an anisotropic object can be calculated for different geometries such as spheres,[29] rigid rods,[34] prolate ellipsoids,[49-51] and double spheres.[52,53] Here we apply both DLS and DDLS to the dumbbell-shaped polymer particles and interpret the experimental data in terms of a slow (Γ_{slow}) and a fast relaxation mode (Γ_{fast}) of the measured intensity autocorrelation function.[26,27,29]

$$\Gamma_{\text{slow}} = D^T q^2 \tag{3.2.1}$$

$$\Gamma_{\text{fast}} = D^T q^2 + 6D^R \tag{3.2.2}$$

For the particles under consideration, D^R characterizes the rotational diffusion around the axis perpendicular to the long symmetry axis of the particles.
We first discuss the results obtained for the dumbbell-shaped PMMA/PS core particles. The experiments were carried out in water and a 10 mM NaCl solution in order to screen the charges on the particle surface. The temperature was set to 25 °C, and the solvent viscosity was 0.891 cp. We used the CONTIN-2DP software to calculate the relaxation frequencies of the different modes from the intensity correlation functions according to eqs. 3.2.1 and 3.2.2. The decay time distributions, that is, the inverse relaxation time distributions, from the DDLS experiments for the dumbbell-particles are shown in Figure 3.2.2 for various scattering angles θ. The peak at the smaller relaxation time, that is, the larger relaxation rate, contains information about the rotational relaxation and the q-dependent part of the translational diffusion (fast mode). The peak at larger relaxation times is due to the translational motion of the particles (slow mode) and should not be seen in a DDLS experiment. However, the slow DDLS mode is visible because of the limited extinction ratio of the Glan-Thompson polarizer (10^{-5}).[54] Moreover, the dispersion of dumbbells is not index matched, which leads to a strong polarized signal.[27,29]

Figure 3.2.3 shows the calculated relaxation rates Γ as a function of q^2 for the PMMA/PS core particles. The slow mode from the DLS experiment (diamonds) must be assigned to the translational motion of the particle, the fast mode from the DDLS experiment can be related to the rotation of the particle (squares). As discussed above, the slow DDLS mode (open circles) can be explained by leakage of the polarizer. As can be clearly seen, this mode is identical with the slow DLS mode (translation) within the experimental error. The fast DLS mode signal (open triangles) is identical with the fast DDLS mode (squares) and describes the rotational relaxation as well as the q-dependent translational term according to eq. 3.2.2. The relaxation rates of the slow mode were fitted according to eq 3.2.1. The data referring to the fast mode were fitted according to eq 3.2.2 only up to a scattering angle of 60° to ensure a reliable resolution of the relaxation (see Figure 3.2.2).

From Figure 3.2.3, D^T was calculated as $(3.29 \pm 0.17) \times 10^{-12}$ m^{2}s^{-1} and D^R as (371 ± 7) s^{-1}. In order to calculate the size parameters of the colloidal particle from the diffusion coefficients, we used the following models: two spheres with radius R_h touching each other (double sphere),[52,53] a prolate ellipsoid with semiaxis a and b ($a > b$)[49-51] and a cylinder with length L and diameter d.[34] Because of the smooth surface of the solid latex particle we assumed stick-boundary conditions.[55] For the viscosity of water at 25 °C, we used the value $\eta = 0.891$ cp (see ref[42]). The calculated size parameters are listed in Table 3.2.1. For all applied geometries, the ratio of the larger to the smaller size parameter is around 2. Evidently, the dumbbell-shaped core particles can be best described as double spheres. Furthermore the size parameters are in good agreement with electron microscopy data, which indicate diameters of (205 ± 5) nm along and (130 ± 7) nm perpendicular to the main particle axis, respectively (see Figure 3.2.1).

Immersing the particles in a 10 mM NaCl solution ($\eta = 0.891$ cp, see ref[43]) did not lead to a significant change in the calculated size parameters within the experimental error as expected.

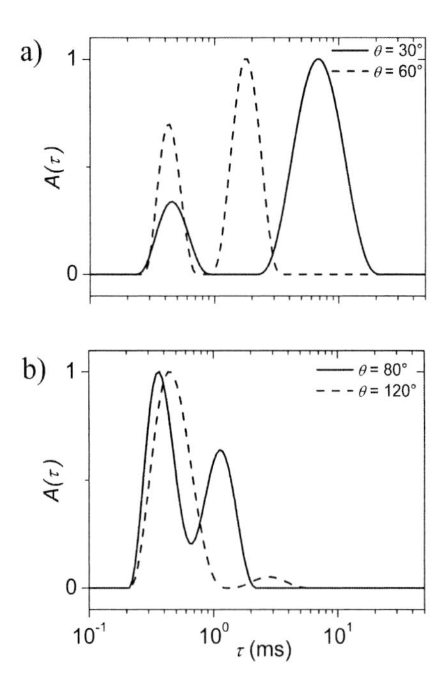

Figure 3.2.2: Relaxation time distributions (CONTIN-plots) as obtained for dumbbell-shaped PMMA/PS latex particles in DDLS experiments (0.1 wt % in water). For each scattering angle, the left peak corresponds to the rotation of the particle and the q-dependent translational term according to eq 3.2.2 (fast mode), the right peak to the translational diffusion according to eq 3.2.1 (slow mode). From the scattering angle $\theta = 30°$ (solid line in panel a) to $\theta = 60°$ (dashed line in panel a), the relaxation processes can be properly resolved by the CONTIN-2DP software. For higher scattering angles the two peaks partly merge: $\theta = 80°$ (solid line in panel b) and $\theta = 120°$ (dashed line in panel b).

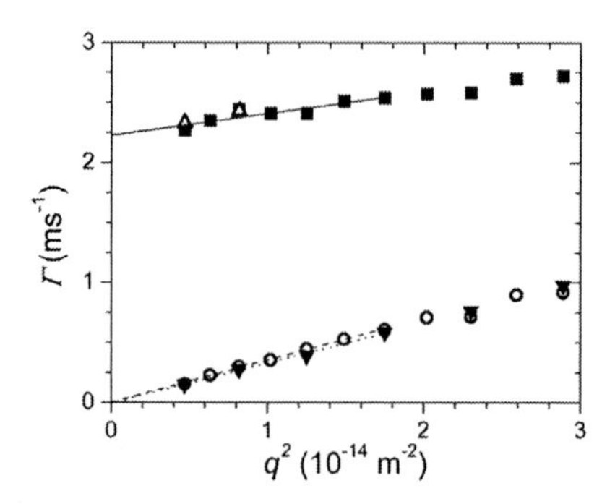

Figure 3.2.3: Relaxation rates as a function of the square of the scattering vector (q^2) for the PMMA/PS dumbbell particles (0.1 wt% in water). Both the rotational relaxation together with the q-dependent translational term (fast mode) and the pure translational term (slow mode) were detected in the DLS and the DDLS experiment. The meaning of the symbols is the following: DDLS fast mode (squares), DLS fast mode (open triangles), DLS slow mode (diamonds), and DDLS slow mode (open circles).

Table 3.2.1: Size Parameters of the Dumbbell-like Core Particles in Water (0.1 wt %) and a 10 mM NaCl Solution (0.01 wt %) Calculated from the Experimental Diffusion Coefficients for Different Geometries under the Assumption of Stick Boundary Conditions.[a]

NaCl [mM]	**geometry**	R_H [b] [nm]	a, b [c] [nm]	L, d [d] [nm]
0	double sphere	51 ± 1		
0	prolate ellipsoid		103 ± 18, 60 ± 10	
0	cylinder			161 ± 7, 103 ± 4
10	double sphere	52 ± 1		
10	prolate ellipsoid		120 ± 14, 48 ± 5	
10	cylinder			189 ± 17, 86 ± 11

[a] In water the translational and rotational diffusion coefficients, $D^T = (3.29 \pm 0.17)10^{-12} \cdot m^2 \cdot s^{-1}$ and $D^R = (371 \pm 7)s^{-1}$, while in a 10 mM NaCl solution $D^T = (3.54 \pm 0.07)10^{-12} \cdot m^2 \cdot s^{-1}$ and $D^R = (348 \pm 8)s^{-1}$. For comparison, electron microscopy data (cryo-TEM) give diameters of (205 ± 20) nm along and (130 ± 7) nm perpendicular to the larger particle axis. For calculations, a viscosity of 0.891 cp for water and the 10 mM NaCl solution at 25 °C was used, respectively. [b] Radius of one sphere. [c] Major (a) and minor (b) semiaxis of a prolate ellipsoid. [d] Length (L) and diameter (d) of a cylinder.

3.2.4.2. Dumbbell-Shaped Polyelectrolyte Brushes

Next, we discuss the synthesis and characterization of the DPB particles. The poly(styrene sulfonate) chains were grafted from the surface of the core particles by photoemulsion polymerization using 2[*p*-(2-hydroxy-2-methylpropiophenone)]-ethylene glycol methacrylate (HMEM) as photoinitiator (Scheme 3.2.1). Shining UV-light on a suspension of these particles leads to the formation of radicals on their surface by photochemical cleavage of the HMEM groups. The monomer sodium styrenesulfonate present in the aqueous phase is thus polymerized to yield linear poly(styrene sulfonate) chains affixed to the core particles. The free 2-hydroxypropyl radicals generated in solution lead to free chains in the aqueous phase which can be separated by ultrafiltration.[1]
The formation of the brush layer was confirmed by the increase of the apparent hydrodynamic radius of the brush particle by means of dynamic light scattering (Figure 3.2.4). The final brush thickness was reached after 40 min which includes a short induction period. Furthermore, as can be seen from the inset of Figure 3.2.4, a low polydispersity for the core (solid line) and the brush particles (dash-dotted line) was found by the CONTIN-analysis[56] of the intensity autocorrelation functions (see inset of Figure 4).

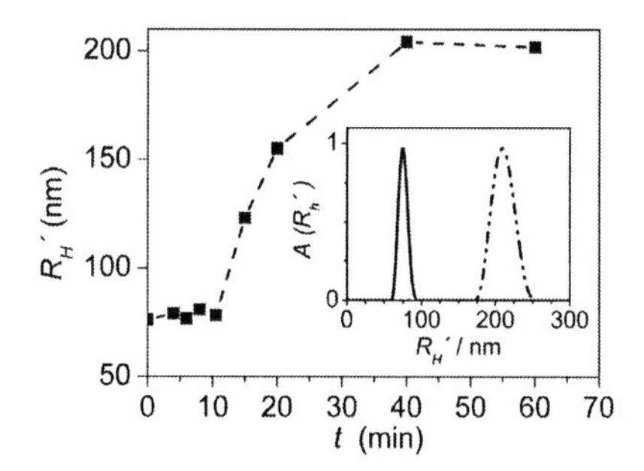

Figure 3.2.4: Increase of hydrodynamic radius (R_H') of the dumbbell brush particles with UV irradiation time. The inset shows the narrow size distributions of the dumbbell-shaped core particles (solid line) and the brush dumbbell particles (dash-dotted line).

Figure 3.2.5 depicts a cryo-TEM image of vitrified dumbbell-shaped brush particles. To increase the contrast, the particles were suspended in a 0.1 mM solution of *N*-dodecylpyridinium chloride (NDPyCl), which forms a polyelectrolyte/surfactant complex with the brush particles.[57] Near the surface of the PMMA/PS cores, about 50 nm of the polyelectrolyte shell is visible. For larger distances the electron density contrast to the background is not sufficient to reveal the brush arms. As shown in the bright field TEM micrograph (inset of Figure 3.2.5), the polyelectrolyte shell collapses on the particle surface if the water is removed.
It is well known that the thickness of polyelectrolyte layers is controlled by the amount

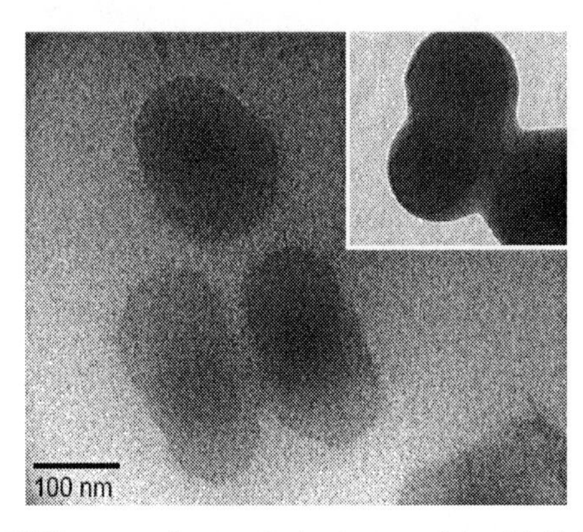

Figure 3.2.5: Cryo-TEM image of a vitrified solution of dumbbell-shaped brush particles (0.2 wt%) in a 0.1 mM NDPyCl-solution. About 50 nm of the brush shell on the particle surface is visible. The inset shows particles in the dried state where the polyelectrolyte shell is collapsed onto the core (TEM, same scale bar).

of salt present in the system.[3,4] A marked shrinking of the hydrodynamic radius of the brush particles for higher ionic strength can be seen in Figure 3.2.6 (open squares), as expected. Hence, the DPB react strongly on an external stimulus which is the concentration of salt in this case. A similar behavior was found in case of the spherical polyelectrolyte brushes recently.[3,4] Compared to sodium chloride, the cationic surfactant *N*-dodecylpyridinium chloride leads to an even stronger shrinking of the brush shell, and thus the hydrodynamic radius (see Figure 3.2.6, filled circles). This is due to the formation of a surfactant/polyelectrolyte complex between the cationic surfactant and the negatively charged polyelectrolyte chains and has been described recently.[57] Because of their surface modification, the DPB particles thus represent a versatile system which is stimuli-responsive for both a broad concentration range of salt as well as surfactants.

3.2.4.3. Rotational Diffusion of DPB Particles

In this section, the results from the DDLS experiments obtained for the DPB particles at various salt concentrations (1 and 100 mM NaCl) are discussed. In addition to the bare dumbbell-shaped cores, each DPB carries a dense shell of anchored polyelectrolyte chains. The chains are almost fully stretched in water and will drastically slow down the diffusive processes, especially at low ionic strength. In contrast to the stiff core particles discussed above, we could resolve three instead of two distinct relaxation modes for scattering angles θ between 20 and 28°. Figure 3.2.7 depicts the results from the CONTIN analysis for DPB in a 1 mM NaCl solution (Figure 3.2.7a) and 100 mM NaCl solution Figure 3.2.7c, respectively. For $\theta = 24°$, the distribution functions show three

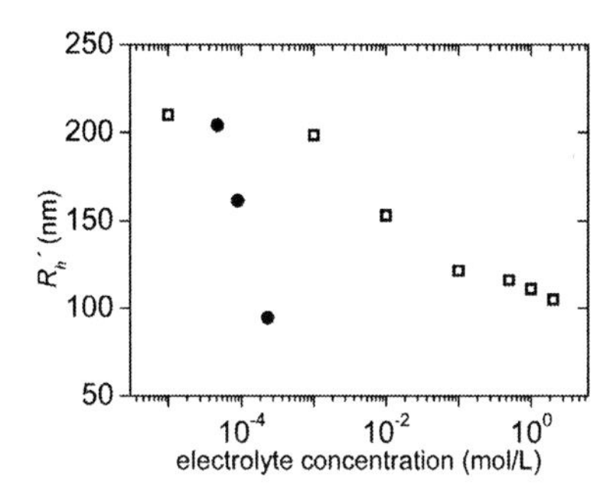

Figure 3.2.6: Dependence of the hydrodynamic radius of the dumbbell-brush particles (0.02 wt%) in the presence of different electrolytes: *N*-dodecylpyridinium chloride (filled circles) leads to a stronger shrinking of the poly(styrene sulfonate) shell compared to a NaCl solution (open squares).

peaks located near 1, 3, and 30 ms (solid lines in Figure 3.2.7a,c). For $28° < \theta < 70°$, the two faster relaxation modes cannot be resolved (dotted lines) and give a single broad peak near 1.5 ms. The slow mode broadens as well. Above $\theta = 70°$ one cannot distinguish the two remaining peaks (dashed lines) as already discussed in the previous section for the core particles. The corresponding relaxation rates Γ were calculated from the relaxation times and are shown as a function of q^2 in Figure 3.2.7b,d, respectively. The translational motion of the particle must be attributed to the open circles (slow DDLS mode), because the linear fit of the data (dashed line) goes through the origin. In addition to the fast mode (squares), which has been connected with the rotation of the dumbbell core particles before (see Figure 3.2.3), another mode comes into play for scattering angles between 20 and 26° (circles), which could not be resolved by the CONTIN analysis for scattering angles larger than 30° (see Figure 3.2.7a,c). This mode will be discussed in the next section.
In the following, we discuss the second mode indicated by squares in Figure 3.2.7b,d. Because there are only a few data points in Figure 3.2.7b,d (squares) that do not suffice to calculate D^R, a linear fit would lead to an enormous experimental error. Thus, we rearranged eq 3.2.2 to

$$D^R = 1/6 \times (\Gamma_{\text{fast}} - \Gamma_{\text{slow}}) \tag{3.2.3}$$

and identified Γ_{fast} as the rotational relaxation frequency and use Γ_{slow} as given in eq 3.2.1. The results in Table 3.2.2 indicate that an increase of salt concentration from 1 to 100 mM leads to a change of the rotational diffusion coefficient from 29 to 67 s^{-1}. We now demonstrate that this mode indicated by squares in Figure 3.2.7b,d is related

to the rotation of the particle by calculating the theoretical thickness of the brush layer from the experimental translational diffusion coefficient (Table 3.2.2). Approximating the core particle by a prolate ellipsoid with the longer semiaxis a = 120 nm and the shorter semiaxis b = 60 nm (see Table 3.2.1), a brush thickness $H_{\text{shell,calc}}$ of 100 nm is obtained for an ionic strength of 1 mM NaCl and of 57 nm for a 100 mM NaCl solution, respectively. This change of the brush thickness by the factor of about 2 is to be expected from earlier studies on spherical polyelectrolyte brushes.[58] Using the resulting brush thickness $H_{\text{shell,calc}}$, theoretical rotational diffusion coefficients D^R_{calc} are obtained which are identical with the experimental values within the experimental error (Table 3.2.2). Within the frame of the double sphere model the diffusion coefficients scale as $D^T \propto 1/(\eta_S R_h)$ and $D^R \propto 1/(\eta_S R_h^3)$, respectively, where η_S is the viscosity of the solvent.
The same scaling holds for spheres and also approximately for cylinders and ellipsoids. Hence the ratio D^T/D^R is proportional to R_H^2 which can be seen from the experimental data, if we set $R_H = R'_h$ and calculate R'_h from the value of D^T using the Stokes-Einstein equation. Hence, we could identify a slow and a fast mode describing the translational and rotational diffusion of the particle in direct analogy to the rigid core particles discussed above.

Table 3.2.2: Experimental Rotational and Translational Diffusion Coefficients D^T and D^R As a Function of Ionic Strength for the DPB particles.[a]

NaCl	D^R	D^T	R_h' [b)]	D^T / D^R	$H_{\text{shell,calc}}$	$D_{\text{calc}}{}^R$	$\Gamma_{\text{coll\_rel}}$
[mM]	[s^{-1}]	[10^{-12} m^2 s^{-1}]	[nm]	[10^{-14} m^2]	[nm]	[s^{-1}]	[s^{-1}]
1	29 ± 8	1.34 ± 0.05	183	4.62 ± 1.28	100	29	≈ 800
100	67 ± 13	1.78 ± 0.05	138	2.66 ± 0.52	57	62	≈ 950

[a] $H_{\text{shell,calc}}$ has been calculated from the D^T value assuming a prolate core particle (a = 120 nm, b = 60 nm). The corresponding D^R_{calc} values are identical with the experimental values D^R within experimental error. $\Gamma_{\text{coll,rel}}$ denotes the frequency for the collective relaxation process detected at small scattering angles. As for the solvent viscosity at 25 °C, we used 0.891 (1 mM NaCl) and 0.898 cp (100 mM NaCl) for all calculations. [b] Hydrodynamic radius calculated from D^T with Stokes-Einstein equation.

3.2.4.4. New Fast Mode: Evidence for Collective Relaxation of the Surface Layer

As discussed above, we observed a second fast mode in a DDLS-experiment for dumbbell-shaped polyelectrolyte brushes which is not related to motion of the particles (see Figure 3.2.7b,d, filled circles). This can be argued directly from the fact that the slow mode and the first fast mode can be assigned to translational and rotational diffusion, respectively. In order to clarify the origins of this mode, we investigated a similar system but

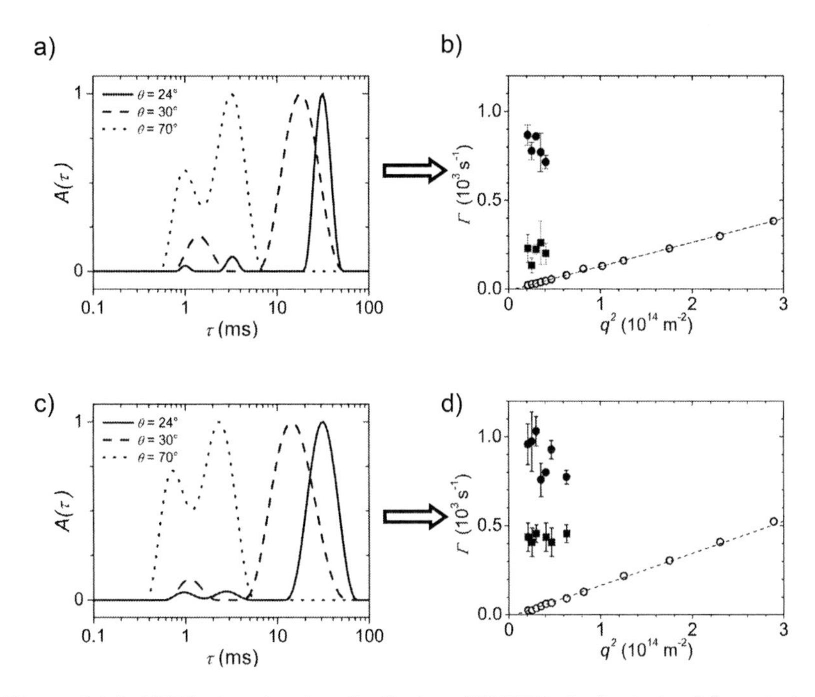

Figure 3.2.7: DDLS-relaxation time distributions (CONTIN-plots) calculated from the intensity autocorrelation functions for the DPB particles in a 1 mM (a) and a 100 mM NaCl solution (c). For the scattering angle $\theta = 24°$, three distinct relaxation modes can be resolved (solid lines) which are connected with the translational motion (slow mode, ≈ 30 ms), the particle rotation (fast mode, $\approx 2-3$ ms), and a collective relaxation process (fast mode, ≈ 1 ms). For $\theta = 30°$ the two faster modes overlap and cannot be resolved (dashed lines). (b,d) The corresponding relaxation rates Γ are plotted as a function of the square of the scattering vector (q^2). Here the translational motion (slow mode) is assigned by open circles, the rotation by squares, and the collective relaxation by circles.

with a spherical core. The synthesis and the characterization of these particles has been described in the Experimental Section. The resulting spherical polyelectrolyte brushes have a core radius of 111 nm and a brush thickness of 89 nm in water. In principle, isotropic spheres do not give rise to a DDLS signal.[26] Figure 3.2.8 shows the same analysis for the spherical polyelectrolyte brushes as given in Figure 3.2.7 for the DPB. Again we observed three modes: the translational and the rotational diffusion and a second fast mode with an relaxation rate $\Gamma_{\text{coll,rel}}$. The diffusion coefficients are listed in Table 3.2.3. It is noteworthy that the relaxation rate $\Gamma_{\text{coll,rel}}$ increases with increasing salt concentration just as in the case of the nonspherical DPB particles (see Table 3.2.2) and has a very similar frequency (≈1000 s^{-1}). This leads to the conclusion that the new relaxation mode found for the DPB particles is independent of the overall particle geometry.

Table 3.2.3: Experimental Diffusion Coefficients D^T and D^R As a Function of Ionic Strength for the SPB Particles.[a]

NaCl [mM]	D^T [10^{-12} m^2 s^{-1}]	R_h [b] [nm]	D^R [s^{-1}]	D^R_{calc} [s^{-1}]	D^T / D^R [10^{-14} m^2]	Γ_{coll_rel} [s^{-1}]
1	1.38 ± 0.07	178	33 ± 5	32	4.18 ± 0.67	≈ 900
100	1.52 ± 0.06	156	50 ± 12	47	3.04 ± 0.74	≈ 1100

[a] As well as for the geometric anisotropic DPB particles, the SPB particles exhibit a collective relaxation rate $\Gamma_{coll,rel}$ = 1000 s^{-1}. [b] Hydrodynamic radius calculated from D^T with Stokes-Einstein equation.

We emphasize that neither modeling the fast relaxation mode as an internal mode of an individual semiflexible polymer chain[59,60] nor using the relaxation time of internal chain motions of grafted flexible polymers[61-63] lead to an agreement with the experimental data. For example, the relaxation rates of a chain molecule of a length L = 100 nm and a persistence length l = 5 nm are larger than about Γ = 100 000 s^{-1}, while the experimental relaxation rate $\Gamma_{coll,rel} \approx 1000\,s^{-1}$ is considerably smaller. Moreover, these internal chain modes exhibit a rather pronounced dependence on the end-to-end distance of the chain molecules, whereas $\Gamma_{coll,rel}$ is rather independent of $H_{shell,calc}$ as shown in Tables 3.2.2 and 3.2.3. Hence, this comparison demonstrates clearly that the fast relaxation mode of both the DPB and SPB particles cannot be explained in terms of internal chain motions of individual polyelectrolyte chains.

Evidently, the fast mode must be assigned to a collective relaxation of the surface layer of polyelectrolyte chains. A key feature of the static scattering intensity of polyelectrolyte brush particles is a strong scattering signal that is due to collective fluctuations of the polyelectrolyte layer on the surface on the particles.[10,64,65] This term has been discussed for a number of systems composed of polymeric layers or networks that are attached to colloidal particles such as core-shell microgels.[66,67] In the case of these polymer gels, the corresponding collective dynamic fluctuations have been measured.[68-70] Interestingly, the obtained relaxation rate $\Gamma_{coll,rel} = D_{coll}q^2 \approx 890\ s^{-1}$ characterizing the collective fluctuations of microgels[69,70] at a scattering angle of 20° is comparable with the relaxation rate $\Gamma_{coll,rel}$ measured for the SPB and DPB particles see Tables 3.2.2 and 3.2.3. Hence we conclude that the fast mode is the dynamic counterpart of the static correlation functions measured earlier in the case of the SPB particles.[10,64,65] Moreover, it is worthwhile to compare the results of the present investigation with earlier studies of the dynamics of polystyrene brushes end-grafted to planar surfaces.[71-73] Collective

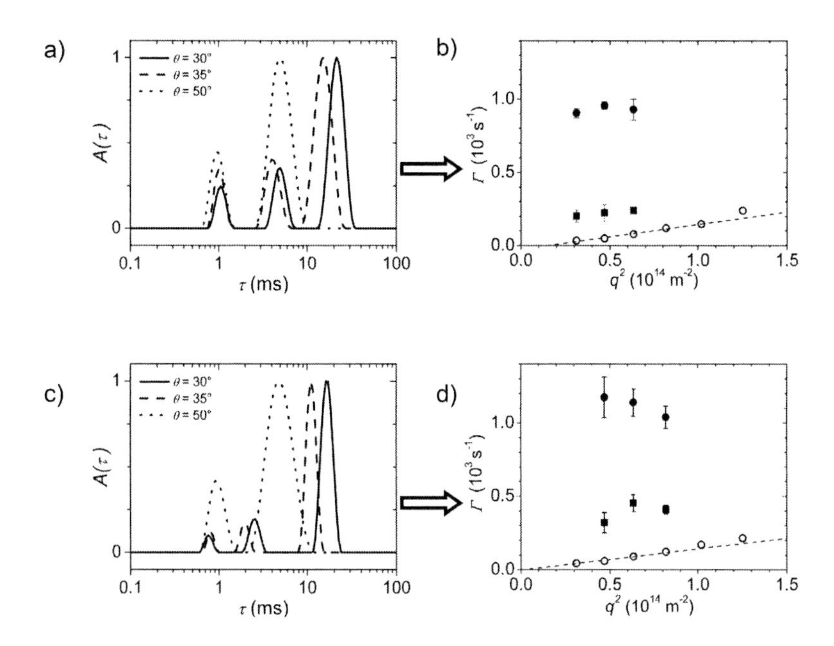

Figure 3.2.8: DDLS-relaxation time distributions (CONTIN-plots) calculated from the intensity autocorrelation functions for the spherical polyelectrolyte brush particles in a 1 mM (a) and a 100 mM NaCl solution (c). Up to the scattering angle $\theta = 35°$, three distinct relaxation modes can be resolved (solid lines and dashed lines) which are connected with the translational motion (slow mode, $\approx 20\,\text{ms}$), the particle rotation (fast mode, $\approx 3\,\text{ms}$), and the collective relaxation process (fast mode, $\approx 1\,\text{ms}$). For $\theta = 50°$, the two faster modes overlap and cannot be resolved (dotted lines). (b,d) The corresponding relaxation rates Γ are shown as a function of the square of the scattering vector (q^2), where the translational motion (slow mode) is signed with open circles, the rotation with squares, and the collective motion with circles.

fluctuations of the grafted polymers have been investigated in terms of a collective relaxation rate $\Gamma_{\text{coll}} = D_{\text{coll}}q^2$ using evanescent wave dynamic light scattering. The resulting calculated relaxation rate $\Gamma_{\text{coll}} \approx 780\ \text{s}^{-1}$ at a scattering angle of $20\,°$ is also comparable to the new mode observed for the SPB and DPB particles. At present, further investigations are conducted in order to clarify this new mode in detail.

3.2.5 Conclusion

In conclusion, we presented the first synthesis and the characterization of colloidal dumbbell-like polyelectrolyte brushes (DPB). The DPB present a stable colloidal system, and the overall dimensions as studied by depolarized dynamic light scattering (DDLS) depend markedly on the concentration of added salt present in the system. In this way the dumbbell-like polyelectrolyte brushes present a novel system of anisotropic colloids that react on external stimuli. For both the spherical and the nonspherical polyelectrolyte brush particles, the DDLS analysis revealed a relaxation mode additional to the rotational and the translational mode that we identified as a collective relaxation of the polyelectrolyte chains.

Acknowledgment

This work was financially supported by the Deutsche Forschungsgemeinschaft, Sonderforschungsbereich 481, Bayreuth. M.H. gratefully acknowledges the Bavarian Elite Network (ENB) Study Program "Macromolecular Science".

3.2.6 References and Notes

[1] Guo, X.; Weiss, A.; Ballauff, M. *Macromolecules* **1999**, *32*, 6043.

[2] Ballauff, M. *Chem. Phys.* **2003**, *204*, 220.

[3] Guo, X.; Ballauff, M. *Langmuir* **2000**, *16*, 8719.

[4] Ballauff, M. *Prog. Polym. Sci.* **2007**, *32*, 1135.

[5] Wittemann, A.; Drechsler, M.; Talmon, Y.; Ballauff, M. *J. Am. Chem. Soc.* **2005**, *127*, 9688.

[6] Mei, Y.; Lu, Y.; Polzer, F.; Ballauff, M.; Drechsler, M. *Chem. Mater.* **2007**, *19*, 1062.

[7] Sharma, G.; Mei, Y.; Lu, Y.; Ballauff, M.; Irrgang, T.; Proch, S.; Kempe, R. *J.Catal.* **2007**, *246*, 10.

[8] Wittemann, A.; Ballauff, M. *Phys. Chem. Chem. Phys.* **2006**, *8*, 5269.

[9] Haupt, B.; Neumann, T.; Wittemann, A.; Ballauff,M. *Biomacromolecules* **2005**, *6*, 948.

[10] Rosenfeldt, S.; Wittemann, A.; Ballauff, M.; Breininger, E.; Bolze, J.; Dingenouts, N. *Phys. Rev. E* **2004**, *70*, 061403.

[11] Henzler, K.; Rosenfeldt, S.; Wittemann, A.; Harnau, L.; Finet, S.; Narayanan, T.; Ballauff, M. *Phys. Rev. Lett.* **2008**, *100*, 158301.

[12] Perro, A.; Reculusa, S.; Ravaine, S.; Bourgeat-Lami, E.; Duguet, E. *J. Mater. Chem.* **2005**, *15*, 3745.

[13] Harnau, L.; Dietrich, S. Inhomogeneous platelet and rodfluids. In *Soft Matter*; Gompper, G., Schick, M., Eds.; Wiley-VCH: Weinheim, 2007; Vol 3, p 159.

[14] Liu, Y.; Abetz, V.; Müller, A. H. E. *Macromolecules* **2003**, *36*, 7894.

[15] Walther, A.; Andre, X.; Drechsler, M.; Abetz, V.; Müller, A. H. E. *J. Am. Chem. Soc.* **2007**, *129*, 6187.

[16] Weber, C. H. M.; Chiche, A.; Krausch, G.; Rosenfeldt, S.; Ballauff, M.; Harnau, L.; Göttker-Schnetmann, I.; Tong, Q.; Mecking, S. *Nano Lett.* **2007**, *7*, 2024.

[17] Manoharan, V. N.; Elsesser, M. T.; Pine, D. J. *Science* **2003**, *301*, 483.

[18] Yi, G.-R.; Manoharan, V. N.; Michel, E.; Elsesser, M. T.; Yang, S.-M.; Pine, D. J. *Adv. Mater.* **2004**, *16*, 1204.

[19] Cho, Y.-S.; Yi, G.-R.; Kim, S.-H.; Jeon, S.-J.; Elsesser, M. T.; Yu, H.K.; Yang, S.-M.; Pine, D. J. *Chem. Mater.* **2007**, *19*, 3183.

[20] Sacanna, S.; Rossi, L.; Kuipers, B. W. M.; Philipse, A. P. *Langmuir* **2006**, *22*, 1822.

[21] Reculusa, S.; Poncet-Legrand, C.; Perro, A.; Duguet, E.; Bourgeat- Lami, E.; Mingotaud, C.; Ravaine, S. *Chem. Mater.* **2005**, *17*, 3338.

[22] Johnson, P. M.; van Kats, C. M.; van Blaaderen, A. *Langmuir* 2005, *21*, 11510.

[23] Sheu, H. R.; El-Aasser, M. S.; Vanderhoff, J. W. *J. Polym. Sci., Part A: Polym. Chem.* **1990**, *28*, 653.

[24] Kim, J.-W.; Larsen, R. J.; Weitz, D. A. *J. Am. Chem. Soc.* **2006**, *128*, 14374.

[25] Mock, E. B.; De Bruyn, H.; Hawkett, B. S.; Gilbert, R. G.; Zukoski, C. F. *Langmuir* **2006**, *22*, 4037.

[26] Berne, B. J.; Pecora, R. *Dynamic light scattering: with Applications to Chemistry, Biology, and Physics*; Dover: NewYork, 2000; pp. 114-129.

[27] Piazza, R.; Stavans, J.; Bellini, T.; Degiorgio, V. *Opt. Commun.* **1989**, *73*, 263.

[28] Matsuoka, H.; Morikawa, H.; Yamaoka, H. *Colloids Surf. A*: **1996**, *109*, 137.

[29] Degiorgio, V.; Piazza, R.; Corti, M.; Stavans, J. *J. Chem. Soc., Faraday Trans.* **1991**, *87*, 431.

[30] Koenderink, G. H.; Philipse, A. P. *Langmuir* **2000**, *16*, 5631.

[31] Koenderink, G. H.; Sacanna, S.; Aarts, D. G. A. L.; Philipse, A. P. *Phys. Rev. E.* **2004**, *69*, 021804.

[32] Eimer, W.; Pecora, R. *J. Chem. Phys.* **1991**, *94*, 2324.

[33] Liu, H.; Skibinska, L.; Gapinski, J.; Patkowski, A.; Fischer, E. W.; Pecora, R. *J. Chem. Phys.* **1998**, *109*, 7556.

[34] Tirado, M. M.; Lopez Martinez, M. C.; Garcia de la Torre, J. *Chem. Phys.* **1984**, *81*, 2047.

[35] Lehner, D.; Lindner, H.; Glatter, O. *Langmuir* **2000**, *16*, 1689.

[36] De Souza Lima, M. M.; Wong, J. T.; Paillet, M.; Borsali, R.; Pecora, R. *Langmuir* **2003**, *19*, 24.

[37] Bica, C. I. D.; Borsali, R.; Rochas, C.; Geissler, E. *Macromolecules* **2006**, *39*, 3622.

[38] Phalakornkul, J. K.; Gast, A. P.; Pecora, R. *J. Chem. Phys.* **2000**, *112*, 6487.

[39] Eimer, W.; Dorfmüller, T. *J. Phys. Chem.* **1992**, *96*, 6790.

[40] Kroeger, A.; Belack, J.; Larsen, A.; Fytas, G.; Wegner, G. *Macromolecules* **2006**, *39*, 7098.

[41] Kim, J. H.; Chainey, M.; El-Aasser, M. S.; Vanderhoff, J. W. *J. Polym. Sci., Part A: Polym. Chem.* **1992**, *30*, 171.

[42] *CRC Handbook of Chemistry and Physics*, 76th ed.; Lide, D. R., Ed.; CRC Press: Boca Raton, Fl, 1995; Chapter 6, p 10.

[43] Hai-Lang, Z.; Shi-Jun, H. *J. Chem. Eng. Data* **1996**, *41*, 516.

[44] Chen, Y.- C.; Dimonie, V.; El-Aasser, M. S. *J. Appl. Polym. Sci.* **1991**, *42*, 1049.

[45] Stubbs, J. M.; Sundberg, D. C. *J. Appl. Polym. Sci.* **2004**, *91*, 1538.

[46] Stubbs, J. M.; Karlsson, O.; Jönsson, J.- E.; Sundberg, E.; Durant, Y.; Sundberg, D. C. *Colloids Surf., A* **1999**, *153*, 255.

[47] Grubb, D. T. *J. Mater. Sci.* **1974**, *9*, 1715.

[48] Ott, H.; Abetz, V.; Altstädt, V.; Thomann, Y.; Pfau, A. *J. Microsc.* **2002**, *205*, 106.

[49] Perrin, F. *J. Phys. Rad.* **1934**, *5*, 497.

[50] Perrin, F. *J. Phys. Rad.* **1936**, *7*,1.

[51] Koenig, S. H. *Biopolymers* **1975**, *14*, 2421.

[52] Carrasco, B.; Garcia de la Torre, J. *Biophys. J.* **1999**, *75*, 3044.

[53] Garcia de la Torre, J.; del Rio Echenique, G.; Ortega, A. *J. Phys. Chem. B* **2007**, *111*, 955.

[54] King, R. J.; Talim, S. P. *J. Phys. E: Sci. Instrum.* **1971**, *4*, 93.

[55] Eimer, W.; Williamson, J. R.; Boxer, S. G.; Pecora, R. *Biochemistry* **1990**, *29*, 799.

[56] Provencher, S. W. *Comput. Phys. Commun.* **1982**, *27*, 213.

[57] Samokhina, L.; Schrinner, M.; Ballauff, M.; Drechsler, M. *Langmuir* **2007**, *23*, 3615.

[58] Mei, Y.; Ballauff, M. *Eur. Phys. J. E* **2005**, *16*, 341.

[59] (a) Harnau, L.; Winkler, R. G.; Reineker, P. *J. Chem. Phys.* **1995**, *102*, 7750. (b) Harnau, L.; Winkler, R. G.; Reineker, P. *J. Chem. Phys.* **1996**, *104*, 6355.

[60] Harnau, L.; Winkler, R. G.; Reineker, P. *Phys. Rev . Lett.* **1999**, *82*, 2408.

[61] Halperin, A.; Alexander, S. *Europhys. Lett.* **1988**, *6*, 329.

[62] Murat, M.; Grest, G. S. *Macromolecules* **1989**, *22*, 4054.

[63] Klushin, L. I.; Skvortsov, A. M. *Macromolecules* **1991**, *24*, 1549.

[64] de Robillard, Q.; Guo, X.; Ballauff, M.; Narayanan, T. *Macromolecules* **2000**, *33*, 9109.

[65] Dingenouts, N.; Patel, M.; Rosenfeldt, S.; Pontoni, D.; Narayanan, T.; Ballauff, M. *Macromolecules* **2004**, *37*, 8152.

[66] Dingenouts, N.; Norhausen, Ch.; Ballauff, M. *Macromolecules* **1998**, *31*, 8912.

[67] Seelenmeyer, S.; Deike, I.; Rosenfeldt, S.; Norhausen, Ch.; Dingenouts, N.; Ballauff, M.; Narayanan, T.; Lindner, P. *J. Chem. Phys.* **2001**, *114*, 10471.

[68] Shibayama, M.; Takata, S.; Norisuye, T. *Physica A* **1998**, *249*, 245.

[69] Hellweg, T.; Kratz, K.; Pouget, S.; Eimer, W. *Colloids Surf., A* **2002**, *202*, 223.

[70] Bolisetty, S.; Hoffmann, M.; Harnau, L.; Hellweg, T.; Ballauff, M. *Macromolecules*, submitted.

[71] Fytas, G.; Anastasiadis, S. H.; Seghrouchni, R.; Vlassopoulos, D.; Li, J.; Factor, B. J.; Theobald, W.; Toprakcioglu, C. *Science* **1996**, *274*, 2041.

[72] Yakubov, G. E.; Loppinet, B.; Zhang, H.; Rühe, J.; Sigel, R.; Fytas, G. *Phys. Rev. Lett.* **2004**, *92*, 115501.

[73] Michailidou, V. N.; Loppinet, B.; Vo, D. C.; Prucker, O.; Rühe, J.; Fytas, G. *J. Polym. Sci., Part B: Polym. Phys.* **2006**, *44*, 3590.

3.3 Thermoresponsive Colloidal Molecules

Martin Hoffmann,[a] Miriam Siebenbürger,[b] Ludger Harnau,[cd] Markus Hund,[e] Christoph Hanske,[e] Yan Lu,[b] Claudia S. Wagner,[a] Markus Drechsler,[f] Matthias Ballauff*[b]

[a] *Physikalische Chemie I, Universität Bayreuth, Universitätsstraße 30, 95440 Bayreuth, Germany. E-mail: martin.hoffmann@uni-bayreuth.de; Fax: +49 (0)921 55 2780; Tel: +49(0)921 55 2217*

[b] *F-I2 Soft Matter and Functional Materials, Helmholtz-Zentrum Berlin, Glienicker Str. 100, 14109 Berlin, Germany. E-mail: matthias.ballauff@helmholtz-berlin.de; Fax: +49 (0)30 8062 2308; Tel: +49 (0)30 8062 3071*

[c] *Max-Planck-Institut für Metallforschung, Heisenbergstr. 3, D-70569, Stuttgart, Germany*

[d] *Institut für Theoretische und Theoretische und Angewandte Physik, Universität Stuttgart, Pfaffenwaldring 57, D-70569 Stuttgart, Germany*

[e] *Physikalische Chemie II, Universität Bayeuth, Universitätsstr. 30, 95440 Bayreuth, Germany*

[f] *Macromolecular Chemistry II, University of Bayreuth, 95440 Bayreuth, Germany*

† *Electronic supplementary information (ESI) available: Detailed descriptions of the synthetic procedures, the measurement methods and the DDLS data analysis are included. See DOI:10.1039/c000434k*

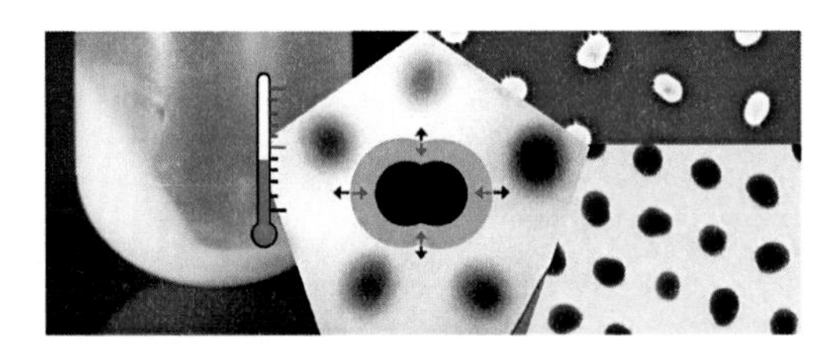

Published in *Soft Matter* **2010**, *6*, 1125-1128.

URL: http://www.rsc.org/Publishing/Journals/SM/article.asp?doi=c000434k

3.3.1 Abstract

We fabricated thermoresponsive colloidal molecules of ca. 250 nm size. Electron- and scanning force microscopy reveal the dumbbell-shaped morphology. The temperature dependence of the size and aspect ratio (ca. 1.4 to 1.6) is analyzed by depolarized dynamic light scattering and found to be in good agreement with microscopic evidence.

Self-assembly of thermoresponsive building blocks is a promising technique to fabricate new materials with tunable properties.[1] Due to their potential applications as biosensors,[2] delivery systems for biomolecules[3] or photonic devices,[4] there is a considerable interest in this fascinating class of material. The most studied example, poly-(*N*-isopropylacrylamide) (PNIPA) undergoes a volume phase transition in water above its lower critical solution temperature (LCST) at ca. 32 °C. Spherical microgels with a solid core and a thermoresponsive shell can be used as colloidal atoms in soft matter physics to mimic transitions from the liquid to the crystalline phase including nonequilibrium phenomena like the glass transition in concentrated suspensions.[5]
However, despite recent advances with PNIPA-coated (spherical) core particles made of silica or gold[6] or raspberry-like stimuli-responsive polymer particles,[7] no attempts have been made for different core geometries so far. Colloidal clusters[8] and anisotropic nanoparticles[9] with shapes resembling space-filling models of simple molecules are expected to show complex behavior like low molecular weight compounds.[10] Thus, extending the analogy between atoms and colloidal spheres, such colloids can be regarded as "colloidal molecules", a term introduced by van Blaaderen.[11] However, the geometry of these particles cannot be changed without chemical modification once synthesized.
In this communication, we describe a strategy that overcomes this limitation by extending our previous work on spherical particles to dumbbell-shaped core-shell microgel particles (DMP).We generate particles with variable morphology by changing the temperature. Field emission scanning electron microscopy (FESEM), cryogenic-transmission electron microscopy (cryo-TEM) and scanning force microscopy (SFM) are used to obtain real space information about the particle morphology. The diffusion coefficients for translational- (D^T) and rotational motion (D^R) are determined by a combination of polarized (DLS) and depolarized dynamic light scattering (DDLS)[12] in the highly diluted regime. Modelling of the diffusion coefficients with the hydrodynamic shell-model gives access to the particle size parameters and the thickness of the PNIPA layer. In this model the particle surface is regarded as a shell of small, non-overlapping and spherical friction elements.[13] It will be shown that the growth of the crosslinked PNIPA shell is not influenced by the core particle geometry. The findings present a significant step towards a general understanding of "colloidal molecules", as in principle the strategy is not limited to a certain particle configuration. Because the synthesis facilitates yields in the gram-scale, the DMP are a versatile model system to investigate the fluid-solid transitions of concentrated dispersions of both thermoresponsive and anisotropic colloids.[14]
We prepared aqueous suspensions of DMP particles with a dumbbell-shaped polymer core and a thermoresponsive PNIPA-shell crosslinked by 5 mol% *N,N'*-methylenebisacrylamide (BIS) (see ESI†for further details of the synthesis). The polymer core particles were synthesized by seeded growth emulsion polymerization using PMMA seeds and styrene monomer added under starved conditions. [12] A relatively high surface tension between the PMMA surface and styrene-monomer as well as the low

reaction temperature (60 °C) and the low concentration of styrene in the water phase (kinetic control) favor the dumbbell morphology (Figure 3.3.1a). After purification by repeated ultracentrifugation, the latex exhibits a red iridescence.
As opposed to previous studies on spherical core-shell microgels,[15] NIPA was not copolymerized with styrene during the formation of the core particles as NIPA is expected to influence the resulting particle morphology. For the formation of the crosslinked thermoresponsive shell, we copolymerized NIPA and BIS (95/5 mol%) in the presence of the PMMA/PS core particles. Due to initiator fragments near the particle surface, the microgels exhibit a zeta-potential of −30.7 mV in water and −2.8 mV in 5 mM KCl at 25 °C. If the microgels are deposited on a weak negatively charged Si wafer, the electrostatic repulsions between the microgels among themselves and with the substrate lead to the 2-dimensional pattern as shown in Figure 3.3.1b (FESEM micrograph). The spreading of the collapsed PNIPA shell on the Si surface is more homogeneous than for negatively charged spherical core-shell microgels as reported recently.[16] Notably, at room temperature the distance between the microgels in the dried state (ca. 250 ± 50 nm) is larger than twice the thickness of the PNIPA layer in the wet state (ca. 100 ± 10 nm) as will be shown by cryo-TEM or DDLS/DLS measurements (see below). This suggests an electrostatic repulsion of the microgels leading to a superlattice when water is evaporated. After centrifugation of diluted DMP solutions, the particles show greenish to bluish iridescence depending on the angle of incident light and observation.
To investigate the particle morphology *in situ*, cryo-TEM measurements[15b,c] were performed (Figure 3.3.1c). The crosslinked PNIPA shell with a thickness of ca. 50 nm (light grey) is clearly visible. Figure 3.3.1c shows that the shell is densely attached to the particle core (dark grey). Due to their statistical orientation in the vitrified water layer, the DMP particles may appear dumbbell-shaped (inset of Figure 3.3.1c) or spherical if the main particle axis is perpendicular to the plane of the micrograph.
Figure 3.3.1d depicts a SFM phase image of the DMP particles arranged on a hydrophilic glass surface in air. The substrate, core particles and the surrounding area can be easily distinguished in the phase image. This indicates that the PNIPA shell is spread around the core particles in accordance with Figure 3.3.1b. Moreover, the electrostatic repulsion between the particles in solution leads to an average distance on the substrate of the order of several diameters.
As demonstrated in recent studies,[8d,12,17] depolarized dynamic light scattering (DDLS) is a highly useful tool to study the hydrodynamics of anisometric particles. DDLS measures the translational (D^T) and the rotational diffusion coefficient (D^R). The comparison of these quantities with hydrodynamic models[12] allows us to gain complementary information of the particle morphology and especially about the thickness of the crosslinked PNIPA layer in aqueous suspension. This information is essential to calculate the volume fraction of the particles in solution for different temperatures. The

analysis done here follows the prescription given recently.[8d,12,17] Thus, the intensity autocorrelation function is measured by DDLS. The slow relaxation rate of the intensity correlation function, $\Gamma_{\text{slow}} = D^T q^2$, was related to translational motion, while the fast decay, $\Gamma_{\text{fast}} = D^T q^2 + 6D^R$, originated from both translational and rotational motion (q: magnitude of the scattering vector). No coupling between rotational motion and shape fluctuations in the PNIPA network has been observed compared to previous studies with spherical core-shell microgels having a lower degree of crosslinking (2.5 mol% BIS) and a thicker PNIPA shell (71.2 ± 2 nm at 25 °C).[17]
As an example of this analysis, Fig. S1 (see ESI†) depicts the fast decay rates Γ_{fast} as a function of q^2 as measured by DDLS (black squares) for the DMP particles at 36.8 °C. The slow relaxation rates from DLS (blue squares) and DDLS experiment (red circles) give the same result for D^T within experimental error. The slow DDLS mode is due to a small leakage of the analyzer.[12] The experimental diffusion coefficients D^T and D^R are shown as a function of the temperature between 14.8 and 36.8 °C in Table 3.3.1. In this range, the shrinking of the crosslinked PNIPA network leads to an increase of D^T by a factor of ca. 2.6 and of D^R by a factor of 5.2. This difference can be understood qualitatively since $D^T \propto R_h^{-1}$, but $D^R \propto R_h^{-3}$ within the framework of the double sphere model,[13a] where R_h is the radius of one constituent sphere.

Table 3.3.1: Diffusion coefficients D^T_{theo} and D^R_{theo} of the DMP particles as calculated with the hydrodynamic shell model together with the results for the particle size R_h and L_h. The experimental diffusion coefficients D^T and D^R are also given for comparison

T (°C)	η (cP)	D^T (10^{-12}m^{2}s^{-1})	D^T_{theo} (10^{-12}m^{2}s^{-1})	D^R (s^{-1})	D^R_{theo} (s^{-1})	R_h (nm)	L_h (nm)
14.8	1.143	1.41 ± 0.04	1.44	60 ± 4	59	110	67
24.5	0.900	2.04 ± 0.03	2.07	98 ± 15	101	99	56
31.8	0.767	2.80 ± 0.04	2.88	193 ± 11	184	83	40
36.8	0.696	3.71 ± 0.02	3.71	304 ± 14	302	70	27

The shell model[13b] for two interpenetrating spheres with radius R_h and a center-to-center distance l can be used to calculate the particle size and shape from the diffusion coefficients. This model is depicted in the inset of Figure 3.3.2. To compare the experimental results with theoretical calculations, we proceeded as follows: from the experimental diffusion coefficients of the PMMA/PS core particles without the PNIPA shell, $D^T_{\text{core}} = (4.14 \pm 0.03) \times 10^{-12}$ m^2 s^{-1} and $D^R_{\text{core}} = (669 \pm 63)$ s^{-1} , the values of the radius $R = 42.7$ nm of the constituent spheres and the center-to-center distance $l = 85.4$ nm between the two spheres of the core were calculated using the shell model for $T = 24.9$ °C and $\eta = 0.893$ cP. Subsequently the thickness of the PNIPA shell L_h (see Figure 3.3.2)

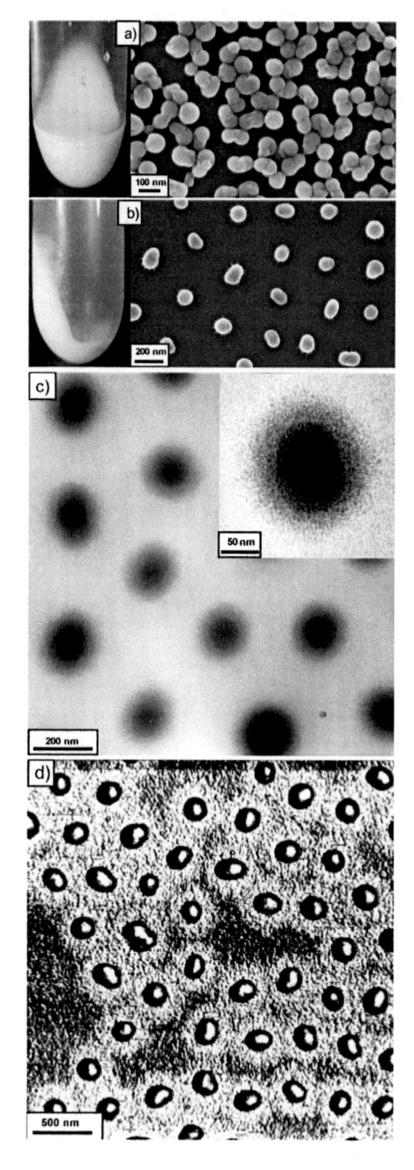

Figure 3.3.1: FESEM micrographs of the a) dumbbell-shaped core particles and b) thermoresponsive core-shell particles (DMP). The collapsed PNIPA shell spreads over a Si wafer. After centrifugation of the suspensions, iridescence can be observed. c) Cryo-TEM micrograph of the core-shell microgel particles. The shell thickness can be estimated as 51.5 ± 5.9 nm for $T = 23$ °C. d) SFM phase image of the DMP particles deposited from a 0.003 wt% solution on a glass slide in air (phase angle $0 - 20\,°$).

was chosen such that the theoretical diffusion coefficients D^T_{theo} and D^R_{theo} match best the experimental values D^T and D^R for each temperature assuming stick-boundary conditions. Table 3.3.1 summarizes the results. With the exception at T = 31.8 °C, the mean deviation between D^T and D^T_{theo} is less than 2% and less than 3% for D^R and D^R_{theo}, respectively. The aspect ratio of the DMP in the observed temperature range is between ca. 1.4 and 1.6. Notably, the shell thickness of the DMP from an *in situ* imaging technique in Figure 3.3.1c of ca. 51.5 ± 5.9 nm is in very good agreement with the precise calculations from the hydrodynamic shell model.

Finally, we demonstrate that the swelling behavior of the PNIPA layer is not influenced by the geometry of the core particles. Figure 3.3.2 shows that the values L_h from Table 3.3.1 are almost identical (±2 nm) to the corresponding data obtained for a spherical reference system[5a] having the same degree of crosslinking (5 mol% BIS). Note that L_h of the reference system was calculated from the translational diffusion coefficient *via* the Stokes-Einstein equation and the core particle radius. This result is of particular importance for the general utility of our strategy towards tuning the shape of colloidal molecules with different morphologies.

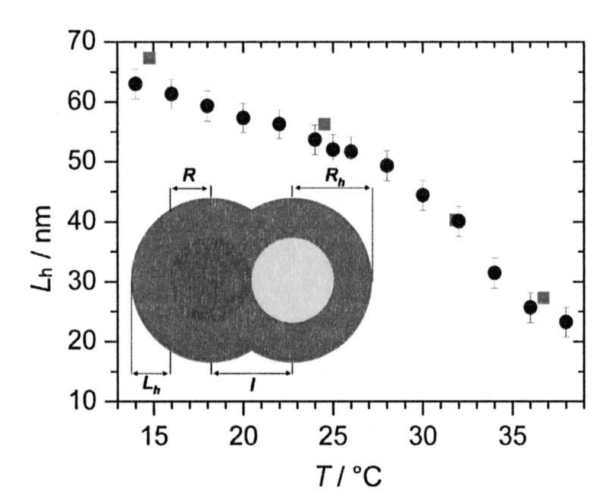

Figure 3.3.2: Shell thickness L_h as a function of temperature determined for the DMP particles with the hydrodynamic shell model (squares) together with the data for a spherical reference system calculated using the Stokes- Einstein equation and the core particle radius (circles). Inset: sketch of the hydrodynamic shell model used for the calculation of the shell thickness $L_h = R_h - R$. The model considers the particle surface composed of small, spherical and non-overlapping friction elements under stick-boundary conditions and allows for an interpenetration of two spheres of radius R_h and a center to center distance l

We have reported here a strategy for the synthesis of thermoresponsive colloidal molecules with a dumbbell-shaped polymer core and a crosslinked shell of poly-(N-

isopropylacrylamide). The procedure used is simple and can be applied to core particles of any structure. The dumbbell-shaped particle morphology was directly proven *in situ* by cryo-TEM. The charge induced 2-dimensional self-assembly of the novel colloids on a Si-substrate was verified by FESEM and SFM. A combined study of conventional and depolarized dynamic light scattering allowed the determination of the temperature dependent PNIPA layer thickness in the whole temperature range using the hydrodynamic shell-model. The microgel particles can change both size (200 to 300 nm) and shape (aspect ratio ca. 1.6 to 1.4) by tuning the temperature between 36.8 and 14.8 °C. Thus, these particles are excellent model systems to study the dynamics of concentrated suspensions of non-spherical colloids.

3.3.2 Notes and references

[1] L. A. Lyon, Z. Meng, N. Singh, C. D. Sorrell and A. St John, *Chem. Soc. Rev.*, 2009, **38**, 865.

[2] S. Su, Md. M. Ali, C. D. M. Filipe, Y. Li and R. Pelton, *Biomacromolecules,* 2008, **9**, 935.

[3] W. H. Blackburn, E. B. Dickerson, M. H. Smith, J. F. McDonald and L. A. Lyon, *Bioconjugate Chem.*, 2009, **20**, 960.

[4] J.-H. Kang, J. H. Moon, S.-K. Lee, S.-G. Park, S.-G. Jang, S. Yang and S.-M. Yang, *Adv. Mater.*, 2008, **20**, 3061.

[5] *(a)* J. J. Crassous, M. Siebenbürger, M. Ballauff, M. Drechsler, O. Heinrich and M. Fuchs, *J. Chem. Phys.*, 2006, **125**, 204906; *(b)* J. J. Crassous, M. Siebenbürger, M. Ballauff, M. Drechsler, D. Hajnal, O. Heinrich and M. Fuchs, *J. Chem. Phys.*, 2008, **128**, 204902.

[6] M. Karg and T. Hellweg, *Curr. Opin. Colloid Interface Sci.*, 2009, **14**, 438.

[7] *(a)* J. Mrkic and B. R. Saunders, *J. Colloid Interface Sci.*, 2000, **222**, 75; *(b)* R. Atkin, M. Bradley and B. Vincent, *Soft Matter,* 2005, **1**, 160.

[8] *(a)* V. N. Manoharan, M. T. Elsesser and D. J. Pine, *Science*, 2003, **301**, 483; *(b)* Y.-S. Cho, G.-R. Yi, S.-H. Kim, S.-J. Jeon, M. T. Elsesser, H. K. Yu, S.-M. Yang and D. J. Pine, *Chem. Mater.*, 2007, **19**, 3183; *(c)* C. S. Wagner, Y. Lu and A. Wittemann, *Langmuir*, 2008, **24**, 12126; *(d)* M. Hoffmann, C. S. Wagner, L. Harnau and A. Wittemann, *ACS Nano*, 2009, **3**, 3326.

[9] *(a)* E. B. Mock, H. D. Bruyn, B. S. Hawkett, R. G. Gilbert and C. F. Zukoski, *Langmuir*, 2006, **22**, 4037; *(b)* J.-W. Kim, R. L. Larsen and D. A. Weitz, *Adv. Mater.*, 2007, **19**, 2005.

[10] *(a)* D. J. Kraft, W. S. Vlug, C. M. van Kats, A. van Blaaderen, A. Imhof and W. K. Kegel, *J. Am. Chem. Soc.*, 2009, **131**, 1182; *(b)* D. J. Kraft, J. Groenewold and W. K. Kegel, *Soft Matter*, 2009, **5**, 3823; *(c)* A. Perro, E. Duguet, O. Lambert, J.-C. Taveau, E. Bourgeat-Lami and S. Ravaine, *Angew. Chem., Int. Ed.*, 2009, **48**, 361; *(d)* C. E. Snyder, M. Ong and D. Velegol, *Soft Matter*, 2009, **5**, 1263.

[11] A. van Blaaderen, *Science*, 2003, **301**, 470.

[12] M. Hoffmann, Y. Lu, M. Schrinner, M. Ballauff and L. Harnau, *J. Phys. Chem. B*, 2008, **112**, 14843.

[13] *(a)* B. Carrasco and J. Garcia de la Torre, *Biophys. J.*, 1999, **76**, 3044; *(b)* J. Garcia de la Torre, G. Del Rio Echenique and A. Ortega, *J. Phys. Chem. B*, 2007, **111**, 55.

[14] *(a)* M. Vega and P. Monson, *J. Chem. Phys.*, 1997, **107**, 2696; *(b)* S. R. Williams and A. P. Philipse, *Phys. Rev. E: Stat. Phys., Plasmas, Fluids, Relat. Interdiscip. Top.*, 2003, **67**, 051301; *(c)* R. Zhang and K. S. Schweizer, *Phys. Rev. E: Stat. Phys., Plasmas, Fluids, Relat. Interdiscip. Top.*, 2009, **80**, 011502.

[15] *(a)* N. Dingenouts, Ch. Norhausen and M. Ballauff, *Macromolecules*, 1998, **31**, 8912; *(b)* M. Ballauff and Y. Lu, *Polymer*, 2007, **48**, 1815; *(c)* J. J. Crassous, C. N. Rochette, A. Wittemann, M. Schrinner, M. Ballauff and M. Drechsler, *Langmuir*, 2009, **25**, 7862.

[16] Y. Lu and M. Drechsler, *Langmuir*, 2009, **25**, 13100.

[17] S. Bolisetty, M. Hoffmann, S. Lekkala, Th. Hellweg, M. Ballauff and L. Harnau, *Macromolecules*, 2009, **42**, 1264.

3.3.3 Electronic Supplementary Information

Thermoresponsive colloidal molecules

Martin Hoffmann,[a] Miriam Siebenbürger,[b] Ludger Harnau,[cd] Markus Hund,[e] Christoph Hanske,[e] Yan Lu,[b] Claudia S. Wagner,[a] Markus Drechsler,[f] Matthias Ballauff*[b]

[a] *Physikalische Chemie I, Universität Bayreuth, Universitätsstraße 30, 95440 Bayreuth, Germany. E-mail: martin.hoffmann@uni-bayreuth.de; Fax: +49 (0)921 55 2780; Tel: +49(0)921 55 2217*

[b] *F-I2 Soft Matter and Functional Materials, Helmholtz-Zentrum Berlin, Glienicker Str. 100, 14109 Berlin, Germany. E-mail: matthias.ballauff@helmholtz-berlin.de; Fax: +49 (0)30 8062 2308; Tel: +49 (0)30 8062 3071*

[c] *Max-Planck-Institut für Metallforschung, Heisenbergstr. 3, D-70569, Stuttgart, Germany*

[d] *Institut für Theoretische und Theoretische und Angewandte Physik, Universität Stuttgart, Pfaffenwaldring 57, D-70569 Stuttgart, Germany*

[e] *Physikalische Chemie II, Universität Bayeuth, Universitätsstr. 30, 95440 Bayreuth, Germany*

[f] *Macromolecular Chemistry II, University of Bayreuth, 95440 Bayreuth, Germany*

ESI 1 Materials

Synthesis of PMMA seed particles

Conventional emulsion polymerization was performed using methyl methacrylate as monomer (MMA; 96.52 g), potassimum peroxodisulfate as initiator ($K_2S_2O_8$; 0.375 g), sodium dodecylbenzene sulfonate (SDBS; 1.499 g) as surfactant and Millipore water as solvent (370.02 g) according to Ref. 1. Ultrafiltration of the diluted suspension (9.126 wt%) was carried out until the conductivity of the eluent was below 4μS cm^{-1}.

Seeded growth polymerization with styrene (core particles)

proceeded as described in Ref. 1 using 78.01 g of the PMMA seed latex (7.12 g solid), styrene (12.258 g), $K_2S_2O_8$ (0.239 g in 14 g water) and additional Millipore water (132.14 g). Monomer and initiator were added under starved conditions (33μL min^{-1}) at 60 °C and stirring at 300 rpm. A uniform particle fraction was obtained with an ultracentrifuge (Allegra 64R, Beckman Coulter; rotor F0650) by repeated centrifugation and redispersion of the diluted suspensions (ca. 4 wt %). First, the very little amount of coagulum formed (larger particles) was removed from the supernatant in runs of

60 min at 4500 rpm and 45 min at 5000 rpm. As the formation of new particles by secondary nucleation could not be fully suppressed, smaller polystyrene particles had to be removed in the supernatant at runs of 75 min at 12000 rpm and 65 min at 12000 rpm.

Polymerization of the crosslinked PNIPA shell

The seeded emulsion polymerization was carried out following the lines in Ref. 2. The latex of the core particles (94.12 g; 4.781 wt %; 4.501 g solid content), NIPA monomer (5.035 g; recrystallized from hexane) and *N,N'*-methylenebisacrylamide (BIS) as a crosslinker (0.327 g) were dissolved together with 12.63 g water in a 250 mL three neck flask equipped with a reflux condenser under nitrogen and stirring at 300 rpm at 23°C for 180 min. After raising the temperature to 80 °C, $K_2S_2O_8$ (0,056 g dissolved in 1.7 g water) was added. The reaction mixture was slowly cooled down to room temperature after 360 min and filtered over glass wool to remove coagulum. The conductivity of the eluent was below $2.4\,\mu S{\cdot}cm^{-1}$ after ultrafiltration of the latex (ca. 4 wt%).

ESI 2 Methods

Dynamic light scattering experiments (DLS/DDLS)

were done with an ALV/DLS/SLS-5000 compact goniometer system (Peters) equipped with a He-Ne laser (632.8 nm). The concentration was 0.003 wt % for the core and 0.002 wt % for the core-shell particles in 5 mM KCl. The samples were filtered through 1 μm nylon filters in dust free quartz glass cuvettes and tempered 120 min to assure conformational equilibrium of the PNIPA network. The cuvettes were immersed in an index matching *cis*-decaline bath. For DDLS (DLS) we measured three runs between 5 and 10 min (1 min) for scattering angles between 20 ° and 42.5 ° (90 °) with angular steps of 2.5 ° (15 °), respectively. The scattered light passed through a Glan Thompson polarizer with an extinction ratio better than 10^{-5}. The relaxation frequencies were obtained by CONTIN-analysis of the intensity autocorrelation functions.

Cryogenic-Transmission Electron Microscopy

Cryo-TEM specimens were prepared by vitrification of thin liquid films (0.2 wt % in Millipore water) supported on a TEM copper grid (Agar G 600HH Cu, Polyscience) in liquid ethane at its freezing point. Examinations were carried out at a Zeiss EM 922Omega EFTEM (Zeiss NTS GmbH, Oberkochen, Germany) at a temperature around 90 K operating at 200 kV. The image was recorded digitally by a bottom-mounted CCD camera system (UltraScan 1000, Gatan) and processed with a digital imaging processing system (Digital Micrograph 3.10 for GMS 1.5, Gatan).

Field Emission Scanning Electron Microscopy

FESEM was done with a LEO 1530 Gemini microscope equipped with a field emission cathode (acceleration voltage 2000 V). The diluted samples (ca. 0.003 wt %) were dried on a Si-wafer in air and sputtered with Pt (Cressington Sputter Coater 208 HR).

Scanning Force Microscopy

SFM measurements were performed with the same samples on glass slides cleaned with the RCA method (see Ref. 3) with a NanoScope Dimension 3100 equipped with a Nanoscope 5 controller and a XYZ hybrid closed-loop scanner. Height and phase images were obtained under ambient conditions (23 °C) by imaging the particles in Tapping™ mode at a frequency of 0.3 Hz (512 samples per line) using an Olympus microcantilever (OMCL-AC160TS).

Electrophoretic mobilities

μ of the particles was measured on a Malvern Zeta-Sizer Nano ZS and converted into the ζ-potential *via* the Smoluchowski equation ($\zeta = \mu\eta\epsilon_0^{-1}\epsilon^{-1}$), where η denotes the viscosity and $\epsilon_0\epsilon$ the relative permittivity of the solution.

ESI 3 DLS/DDLS-analysis

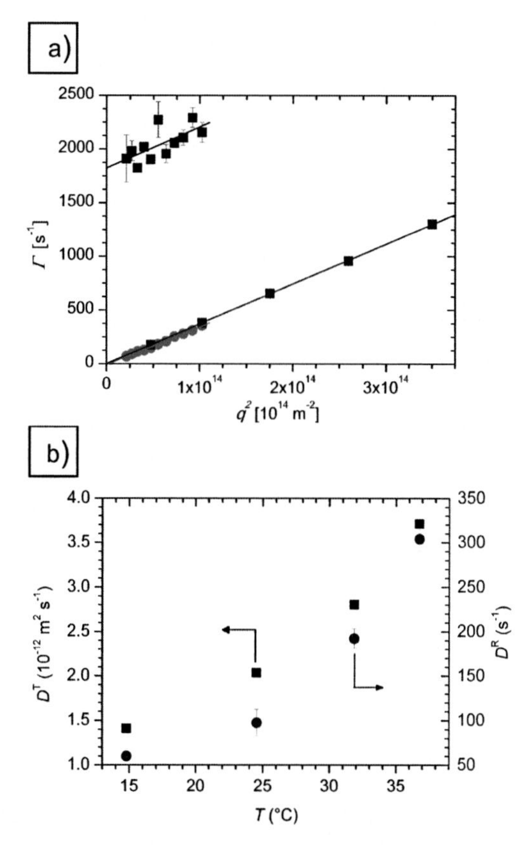

Fig. ESI 1 a) Relaxation rates as a function of the square of the scattering vector q^2 for the DMP particles (0.002 wt % in 5 mM NaCl at 36.8 °C). Both the rotational relaxation together with the q-dependent translational term (fast mode) and the pure translational term (slow mode) were detected in the DLS and the DDLS experiment. DDLS fast mode: black squares, DDLS slow mode: red circles, and DLS slow mode: blue squares. b) Translational (D^T; black squares, left ordinate) and rotational diffusion coefficients (D^R; blue circles) for the DMP core-shell particles (0.002 wt % in 5 mM KCl) at different temperatures as obtained from the DLS/DDLS measurements.

References

(1) M. Hoffmann, Y. Lu, M. Schrinner, M. Ballauff and L. Harnau, *J. Phys. Chem. B.* 2008, **112**, 14843.

(2) N. Dingenouts, Ch. Norhausen and M. Ballauff, *Macromolecules* 1998, **31**, 8912.

(3) W. Kern and D. A. Puotinen. *RCA Review* 1970, **31**, 187.

3.4 3D Brownian Diffusion of Submicron-Sized Particle Clusters

Martin Hoffmann,[†] Claudia S. Wagner,[†] Ludger Harnau[‡,S] and Alexander Wittemann[†*]

[†]*Physikalische Chemie I, Universität Bayreuth, Universitätsstr. 30, 95440 Bayreuth, Germany*

[‡] *Max-Planck-Institut für Metallforschung, Heisenbergstr. 3, D-70569 Stuttgart, Germany*

[S] *Institut für Theoretische und Angewandte Physik, Universität Stuttgart, Pfaffenwaldring 57, D-70569 Stuttgart, Germany*

[*]Corresponding author: alexander.Wittemann@uni-bayreuth.de

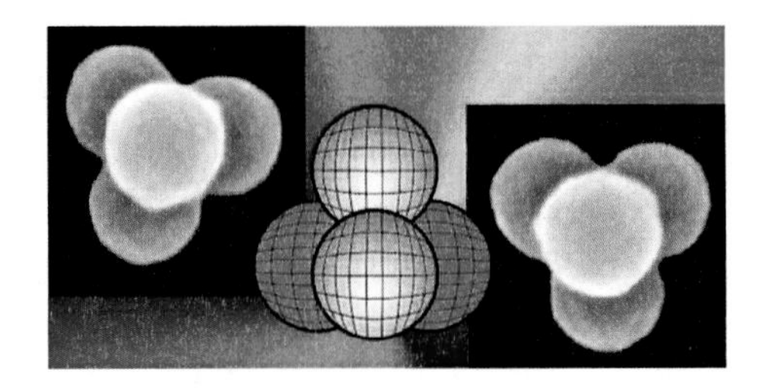

Published in *ACS Nano* **2009**, *3*, 3326-3324.

3.4.1 Abstract

We report on the translation and rotation of particle clusters made through the combination of spherical building blocks. These clusters present ideal model systems to study the motion of objects with complex shape. Since they could be separated into fractions of well-defined configurations on a sufficient scale and because their overall dimensions were below 300 nm, the translational and rotational diffusion coefficients of particle doublets, triplets, and tetrahedrons could be determined by a combination of polarized dynamic light scattering (DLS) and depolarized dynamic light scattering (DDLS). The use of colloidal clusters for DDLS experiments overcomes the limitation of earlier experiments on the diffusion of complex objects near surfaces because the true 3D diffusion can be studied. When the exact geometry of the complex assemblies is known, different hydrodynamic models for calculating the diffusion coefficients for objects with complex shapes could be applied. Because hydrodynamic friction must be restricted to the cluster surface, the so-called shell model, in which the surface is represented as a shell of small friction elements, was most suitable to describe the dynamics. A quantitative comparison of the predictions from theoretical modeling with the results obtained by DDLS showed an excellent agreement between experiment and theory.

KEYWORDS: colloidal clusters, diffusion, Brownian motion, rotation, depolarized dynamic light scattering, shell model, ellipsoids of revolution

3.4.2 Introduction

Translational and rotational diffusion of colloidal particles was extensively studied by experiment,[1-7] simulation, [8] and theory[9-13] over the past decades. In most of these studies, colloidal particles with simple shapes such as spheres,[14-18] ellipsoids,[19,20] rods,[21-26] and platelets[27] were used. The hydrodynamic properties of well-defined model colloids with shapes that differ from these simple geometries are important to understand the diffusion of objects with complex shapes.[28] The dynamics of complex particles are fundamental to the understanding of many practical problems such as biodistribution, sedimentation, coagulation, flotation, and rheology.[29] Hence, Granick and co-workers prepared different planar clusters from micron-sized particles.[30,31] Such particle clusters are ideal candidates for the study of the motion of complex objects because they exhibit well-defined geometries[32] The 2D diffusion of the clusters could be studied by video microscopy because the overall dimensions of the clusters were in the micron range.[30] The planar particle assemblies were prepared through evaporation of a suspension of silica microspheres on a microscope slide. The randomly distributed planar aggregates were cemented together and resuspended in an aqueous solution.[30] This technique is limited to micron-sized planar assemblies and small quantities of clusters. Velev and co-workers prepared 3D assemblies of microspheres using emulsion droplets as a template for the cluster formation.[33,34] Pine, Bibette, and coworkers developed this technique further for the production of clusters having well-defined configurations.[35,36] Packing of the microspheres was accomplished through the agglomeration of the particles adsorbed onto the surface of macroemulsion droplets.[37] Recently, we combined this approach with basic principles established in the miniemulsion technique.[38] Colloidal clusters with overall dimensions below 300 nm could be obtained using narrow-dispersed emulsion droplets with diameters of 1.9 μm prepared through power ultrasonication and monodisperse spherical building blocks with diameters of 154 nm.[38] Clusters of these dimensions underlie Brownian motion, which prevails over gravitational forces. The dynamics of these submicron-sized clusters with well-defined configurations can be thus studied by dynamic light scattering techniques. Hence, they present excellent model systems to study the diffusion of particles with complex shapes. Polarized dynamic light scattering (DLS) became a routine technique to measure translational diffusion coefficients of submicron-sized particles. In DLS, the incident light is usually vertically polarized. The scattered light is dominated by the vertically polarized contribution, but it can contain a horizontally polarized contribution, as well.[22,39] In depolarized dynamic light scattering (DDLS) experiments, the latter contribution to the scattered light is measured through a horizontally oriented polarizer, such as a Glan-Thompson prism.[39] Pecora and coworkers could demonstrate that both translational and rotational diffusion coefficients can be derived from the contribution of the horizontally polarized scattered light.[22]
DDLS has been frequently applied to small molecules[40-42] but less often to large molecules or particles because of their relatively weak depolarized signal.[19,26] However,

different theoretical models and computational procedures have been proposed for the calculation of the hydrodynamic properties of complex particles. A comprehensive overview of model building and hydrodynamic calculation is given in refs.[28, 43, 44].
Here we present for the first time a study of both the translational and rotational diffusion of submicron-sized colloidal clusters consisting of up to four building blocks through a combination of DLS and DDLS. This technique could be used because the clusters underlie Brownian motion. Hence, in contrast to self-diffusion measurements of micron-sized objects by microscopy,[30,45] the true 3D diffusion of complex colloids excluding wall effects can be studied by DDLS. Moreover, different models[28,44] that have been proposed for the description of the hydrodynamic properties of complex particles were probed to predict the translational and rotational diffusion coefficients of the clusters with regard to their configuration. The theoretical results were then compared with the experimental results to get a clear understanding of the hydrodynamics of the complex assemblies.

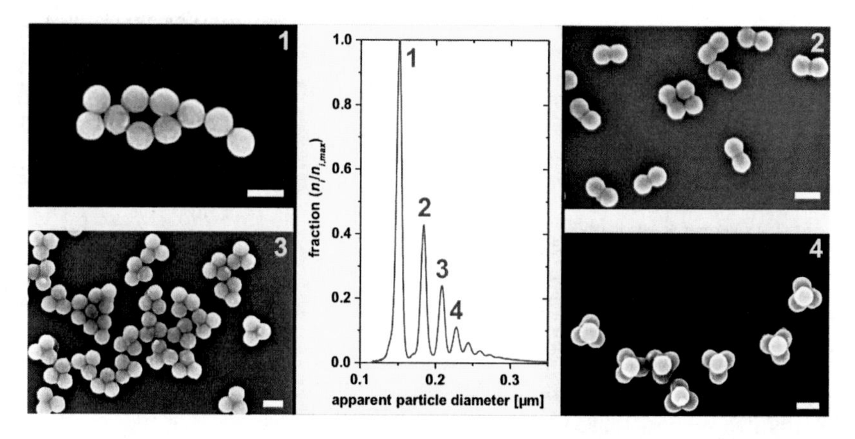

Figure 3.4.1: Sample preparation: The clusters were fabricated through the aggregation of particles adsorbed onto narrow-dispersed emulsion droplets. The statistical distribution of the building blocks onto the droplets led to clusters of different size. The number-weighted size distribution of the clusters obtained by analytical disk centrifugation shows that essentially large amounts of small clusters were formed (center). Because of their different sedimentation velocity, the clusters could be separated through centrifugation in a density gradient into fractions of single particles (1), particle doublets (2), triplets (3), and tetrahedrons (4). The FESEM micrographs demonstrate that suspensions of uniform clusters were obtained which present model systems for particles with complex shapes. Scale bars are 200 nm.

3.4.3 Results and Discussions

3.4.3.1. Particle Clusters

We studied dilute aqueous suspensions of particle clusters with a specific number of constituents N such as particle doublets ($N = 2$), triplets ($N = 3$), and tetrahedrons ($N = 4$). These clusters consisted of amino-modified polystyrene particles of 154 nm in diameter (Figure 3.4.1). The assembly into clusters was accomplished using narrow-dispersed emulsion droplets as templates.[38] The colloidal building blocks were adsorbed onto the oil droplets because the adsorption lowers the interfacial energy due to the Pickering effect. Subsequent evaporation of the oil causes capillary forces which make the particles pack together. This process leads to clusters of well-defined configurations (Figure 3.4.1) which are believed to result from a preorientation of the particles already at the droplet surface.[37,46]
Figure 3.4.1 shows that the suspension of the clusters can be separated by centrifugation into fractions of uniform clusters because of their different sedimentation velocity in a density gradient.[38] The analysis of the dynamics was restricted to clusters made from up to four constituents because these clusters have only one distinct configuration. Cluster consisting of more than four particles may have more than one configuration. For example, five spherical building blocks can be assembled into triangular dipyramids or square pyramids. Moreover, the smaller clusters could be prepared in scales which are sufficient for scattering experiments, and they can be separated into uniform fractions by centrifugation because of the large difference in mass of small assemblies (Figure 3.4.1).

3.4.3.2. DDLS Experiments

In the following, we discuss the DLS and DDLS analysis of the submicron-sized clusters, which were used to determine their translational and rotational diffusion coefficients D^T and D^R, respectively. The principle of the experimental setup is shown in Figure 3.4.2A. In both DLS and DDLS, the incident light was vertically polarized. The DDLS intensity autocorrelation functions (Figure 3.4.2B) presented the sum of two discrete exponential decays, where the slow mode was related to the translational diffusion while the fast mode originated from translational and rotational diffusion. The relaxation rates of the slow and the fast modes, Γ_{slow} and Γ_{fast}, of the autocorrelation functions can be expressed as follows:[4]

$$\Gamma_{\text{slow}} = D^T q^2 \tag{3.4.1}$$

$$\Gamma_{\text{fast}} = 6D^R + D^T q^2 \tag{3.4.2}$$

In DLS, the autocorrelation functions are dominated by the slow mode with the relaxation rate Γ_{slow} (eq 3.4.1). The contribution of the fast mode to the autocorrelation

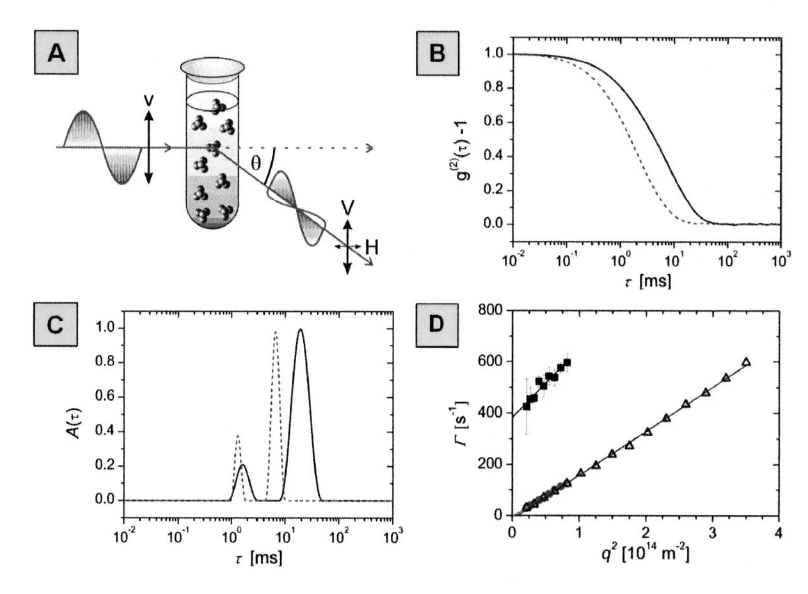

Figure 3.4.2: Depolarized dynamic light scattering (DDLS) experiments of particle tetrahedrons ($N = 4$): (A) schematic representation of the experiment; the incident light is vertically polarized (v). Moreover, the signal of the light scattered by the clusters is mainly governed by vertically polarized light (V), as well. In DDLS, a horizontally oriented polarizer is used to measure the horizontally polarized component (H) of the scattered light. The latter is zero for isotropic particles but nonzero for scatters with optical or shape anisotropy. (B) Depolarized intensity autocorrelation functions ($g^{(2)}(\tau) - 1$) measured at 30° (blue line) and 40° (red dashed line) for a suspension of particle tetrahedrons. (C) Relaxation time distributions $A(\tau)$ (CONTIN plots) as derived from the autocorrelation functions. The right peak (slow relaxation mode) originates from the translational diffusion of the clusters, whereas the left peak (fast relaxation mode) contains information on both the translation and rotation of the complex assemblies. (D) Relaxation rates Γ as the function of the square of the absolute value of the scattering vector q: slow modes of DLS (blue triangles) and DDLS (red spheres); fast mode of DDLS (black spheres). The linear correlations follow eq 3.4.1 (slow modes) and eq 3.4.2 (fast modes), respectively, which in turn give access to the translational and rotational diffusion coefficients.

function is poor and can be neglected as long as the scattering objects have dimensions which are on the same order as the inverse absolute value of the scattering vector $1/q$ ($q = 4\pi n/\lambda \sin(\theta/2)$), where n is the refractive index of the solvent, λ the wavelength, and θ the scattering angle).

This is the case for the clusters with dimensions below 300 nm. Hence, DLS could be used to study the translational diffusion of the clusters but not for the analysis of their rotational dynamics. Therefore, we used DDLS to study the rotation of the clusters, as well.

In DDLS, the horizontally polarized component of the light scattered by the clusters is detected through a polarizer (vH detection).[39] This contribution to the scattered light is much smaller than the intensity of the vertically polarized scattered light (vV detection) measured in a DLS experiment. This did not present an obstacle for the study of the cluster hydrodynamics because the concentrations of the cluster suspensions were still in the dilute regime but high enough to record intensity autocorrelation functions in vH detection (Figure 3.4.2B). In principle, the depolarized autocorrelation functions of the clusters should be characterized by a single exponential decay with the relaxation rate Γ_{fast} following eq 3.4.2. However, the autocorrelation functions were the sum of two discrete experimental decays. This was further corroborated by CONTIN analysis[47] of the autocorrelation functions, which was used to calculate the distribution of the relaxation times $A(\tau)$. In all cases, bimodal distributions of the relaxation times τ were obtained (Figure 3.4.2C). The fast mode originated from translational and rotational diffusion of the clusters according to eq 3.4.2 (Figure 3.4.2D; see also Supporting Information Figures 1 and 2). Hence, it can be used to obtain D^T and D^R. The second mode, that is, the slow mode, emerges from vertically polarized scattered light as a consequence of the limited extinction ratio of the Glan-Thomson polarizer (10^{-5}). Because this mode follows eq 3.4.1, it provides an additional access to D^T (Figure 3.4.2D).

The particle clusters are bearing amino groups on their surface, which leads to electrostatic stabilization. Electrostatic repulsion among the clusters might affect the dynamics.[48,49] Hence, we immersed the clusters in solutions of 10 mM NaCl to screen the charges. The results did not differ from the values of D^T and D^R measured in pure water. Hence, all further experiments could be performed in water because the electrostatics did not influence the dynamics of the clusters in the dilute regime.

The building blocks of the clusters should be optically isotropic because of their spherical shape. However, even for these particles, a depolarized signal could be measured. Despite the low polydispersity of the building blocks of the clusters, there might be slight deviations from either the spherical shape or an uneven distribution of the amino groups on the surface, which makes the particles optically anisotropic. The depolarized signal caused by such effects is rather poor. At a scattering angle of 40°, the contribution of the fast mode to the intensity of the scattered light is only 3 %. Nonetheless, it could be used to determine D^R in addition to D^T (see Supporting Information Figure 1). Of course, D^R is affected by a larger error in this case than for the particle doublets and triplets which have an anisotropic shape. The shape anisotropy of the doublets and triplets results in large depolarized signals, that is, 30 and 21 %, respectively, of the total intensity of the scattered light. The intensity of the depolarized signal increases with the shape anisotropy of the assemblies (see Supporting Information Figure 3). Therefore,

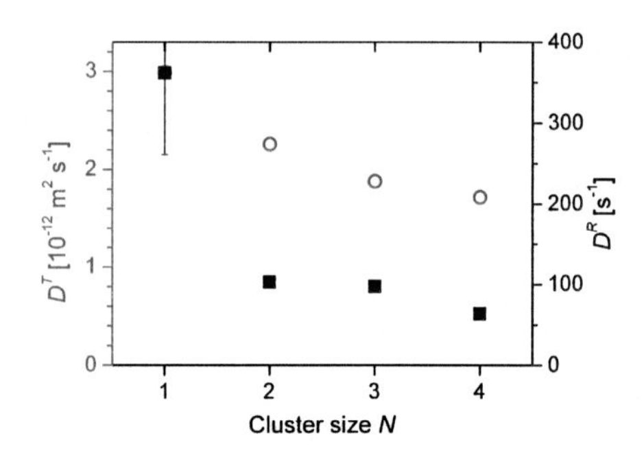

Figure 3.4.3: Translational diffusion coefficients D^T (red circles) and rotational diffusion coefficients D^R (blue squares) of the particle clusters as the function of the number of building blocks N (single particles, $N = 1$; particle doublets, $N = 2$; triplets, $N = 3$; tetrahedrons, $N = 4$).

DDLS is a suitable method to especially study the dynamics of dumbbell-shaped[3] or rod-like particles.[26]

According to eqs. 3.4.1 and 3.4.2, the linear regressions of Γ_{fast} and Γ_{slow} on q^2 shown in Figure 3.4.2D should have the same slopes, that is, the translational diffusion coefficient D^T. This holds as long as the translation and rotation motions of the clusters are decoupled. We observed a perfect agreement of D^T as obtained from DLS and from both modes for the particle doublets and triplets, which in turn confirms the decoupling of rotational from translational diffusion (see Supporting Information Figure 2). For the tetrahedrons, the slopes of the linear regressions shown in Figure 3.4.2D slightly differ. Because of the low volume fractions of the cluster suspensions of less than 10^{-5}, interparticular interactions and a coupling between rotation and diffusion can be excluded. [50] The deviation might be due to the larger experimental error of the DDLS measurement of tetrahedrons because a tetrahedron has rather small shape anisotropy as compared to a doublet and a triplet. However, the building blocks are slightly optically anisotropic. Moreover, they may also slightly differ in size, which contributes to the shape anisotropy. For tetrahedrons, the fast mode contributes 10 % to the total scattering intensity of the DDLS experiment.

Figure 3.4.3 shows the translational and rotational diffusion coefficients of the different cluster species as derived from the Γ *versus* q^2 plot according to eqs. 3.4.1 and 3.4.2. D^T decreases smoothly with increasing size of the clusters. Translational diffusion is widely dominated by the volume of the assemblies, whereas the shape plays a minor role. Hence, the decrease of D^T with the cluster size follows widely from the increase in the mean radius with the number of constituent particles of the clusters. However, the rotational characteristics differ from the translational properties. When going from

the single particle to the particle doublet, a marked drop of D^R is observed, whereas the particle triplet has almost the same rotational diffusion coefficient as the doublet (Figure 3.4.3). Similar observations were made for micron-sized clusters.[30]
D^R decreases again by 2/3 when going to the tetrahedron. Hence, it is evident that the decrease of D^R does not exclusively correlate with the cluster volume because the rotational properties are strongly determined by the shape of the clusters, as well. To gain further insight into the experimental data, we use two different models which will be discussed in the following sections.

3.4.3.3. Simple Description of the Dynamics on the Basis of Spheroids

As mentioned above, the dynamics of the rigid clusters in solution encompass translational and rotational motion. These motions correlate directly with the overall size and shape of the clusters. The solvent is assumed to obey the low Reynolds number Navier-Stokes equation and incompressibility equations.[43] Hence, the hydrodynamic properties of a rigid object are contained in a diffusion matrix that provides a linear relationship between velocities and angular velocities to forces and torques acting on the body. In principle, the diffusion matrix can be calculated by solving the Navier-Stokes equation, but this is usually prohibitively difficult due to the complex shape of the objects. However, spherical bodies and spheroidal shapes are among the few shapes for which the flow equations of hydrodynamics can be solved exactly. For ellipsoids of revolution with two semiaxes of equal length, the translational ($D_\parallel^T, D_\perp^T$) and rotational ($D_\parallel^R, D_\perp^R$) diffusion coefficients parallel and perpendicular to the main symmetry axis are given by[43]

$$D_\parallel^T = \frac{k_B T}{8\pi\eta a}\frac{(2-p^2)G(p)-1}{1-p^2} \tag{3.4.3}$$

$$D_\perp^T = \frac{k_B T}{16\pi\eta a}\frac{(2-3p^2)G(p)+1}{1-p^2} \tag{3.4.4}$$

$$D_\parallel^R = \frac{3k_B T}{16\pi\eta a^3 p^2}\frac{1-p^2G(p)}{1-p^2} \tag{3.4.5}$$

$$D_\perp^R = \frac{3k_B T}{16\pi\eta a^3}\frac{(2-p^2)G(p)-1}{1-p^4} \tag{3.4.6}$$

where

$$G(p) = \log\left(\frac{1+\sqrt{1-p^2}}{p}\right)/\sqrt{1-p^2}$$

for $p < 1$ and

$$G(p) = \arctan(\sqrt{p^2-1})/\sqrt{p^2-1}$$

for $p > 1$.
Here a is the semiaxis along the axis of revolution, b is the equatorial semiaxes, and $p = b/a$ is the axial ratio. In the case of prolate ellipsoids, the axial ratio p is smaller than 1 since the axial semiaxis a is longer than the equatorial semiaxes b. Conversely, $p > 1$ in oblate ellipsoids because the axial semiaxis a is shorter than the equatorial semiaxes b. Finally, spheres have an axial ratio of 1 because all three semiaxes are equal in length. The orientation-averaged translational diffusion coefficient can be expressed as

$$D^T = \frac{D_{\parallel}^T + 2D_{\perp}^T}{3} \tag{3.4.7}$$

D^T can be measured by DLS and DDLS according to eqs 3.4.1 and 3.4.2. The rotational diffusion around the axis of revolution characterized by $D_{\parallel}^R$ can be detected provided the ellipsoidal particles exhibit an optical anisotropy of sufficient magnitude. Otherwise, only the rotational diffusion coefficient $D_{\perp}^R$ can be measured by DDLS due to the particle shape anisotropy.
As a first approximation, one may use eqs 3.4.3- 3.4.7 in order to model the diffusion coefficients of the clusters under consideration. Using the temperature $T = 298.15$ K and the solvent viscosity $\eta = 0.000891$ N s m^{-2} as input into eqs 3.4.3- 3.4.7 leads to the diffusion coefficients given in Figure 3.4.4. These calculated diffusion coefficients are similar to the experimental data shown in the left column of Figure 3.4.4. Hence, the dynamics of the clusters can be described in terms of the dynamics of a sphere for $N = 1$ and 4, a prolate ellipsoid for $N = 2$, and an oblate ellipsoid for $N = 3$, where N is the number of the building blocks. However, the length of the semiaxis a and the axial ratio p cannot be determined directly from the size and shape of the particle clusters in the case of $N = 2$, 3, and 4, but these parameters follow from modeling the experimental data with the help of eqs 3.4.3- 3.4.7. Thus, this model gives already a first description of the diffusion of the clusters. However, because its parameters cannot be derived directly from the geometry of the clusters, hydrodynamic models were probed that allow incorporating the true shape.

3.4.3.4. Modeling of the Cluster Dynamics Based on Hydrodynamic Models

In order to take into account the shape of the clusters correctly, we use the so-called shell model, in which the surface of the clusters is covered with a large number of nonoverlapping spherical friction elements. This model is in widespread use, and public-domain computer programs are available (see, *e.g.*, refs.[28, 44, and 51] and references therein). The hydrodynamic interaction between the beads is the crucial point in the numerical computation of the diffusion coefficients. Replacing a complex particle surface by a shell of very small spherical friction elements will give the correct diffusion coefficients, provided the diffusion matrix can be calculated numerically for a large number of small beads.

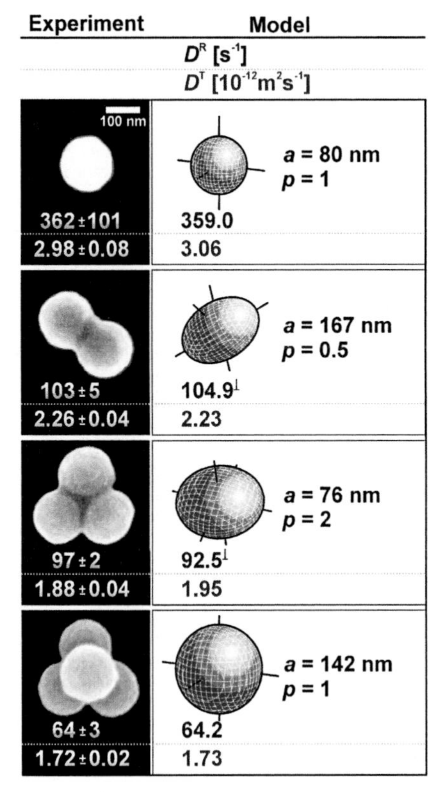

Figure 3.4.4: Translational (D^T) and rotational (D^R) diffusion coefficients as obtained by DLS and DDLS, respectively, together with the diffusion coefficients calculated for spheroids according to eqs. 3.4.3- 3.4.7. The length of the axial semiaxis a of the spheroids and the axial ratio p are input estimates which cannot be directly correlated with geometric parameters of the clusters. For the particle doublets and triplets as well as for the corresponding ellipsoids, $D^R_\perp$ perpendicular to the main axis of symmetry is measured and calculated, respectively. The SEM micrographs show the particle clusters oriented with their main body parallel to the plane of the figure (left column).

Figure 3.4.5 presents the calculated translational and rotational diffusion coefficients together with the experimental results. In the calculations, the radius $R_S = 80$ nm of the constituent spheres and the center-to-center distance $L = 145$ nm between two spheres within a cluster were used. L is taken directly from the field emission scanning electron microscopy (FESEM) images shown in Figure 3.4.1. It differs from the diameter of the building blocks because the clusters do not consist of touching constituent spheres. This is due to the assembly of the building blocks from toluene droplets. At the oil-water interface, the crosslinked polystyrene particles are partly swollen in toluene because toluene is a good solvent for polystyrene. Capillary forces created by the evaporation of the toluene pack the plasticized particles together in their final configuration. This leads to a partial deformation of the spheres and enhances the contact area among the constituent spheres within the cluster.
From the comparison of the experimental data with the calculated results, it is apparent that the shell model is an appropriate theoretical tool for studying the dynamics of these systems (Figure 3.4.5). In addition, we used a bead model[28,44] in which the particle clusters are represented directly by big spheres of radius R_S. Using the same model parameters as for the shell model leads to translational and rotational diffusion coefficients which differ by less than 3 % from the results of the shell model calculations. Hence, the bead model confirms the results obtained with the shell model.
We note that the experimental results cannot be explained assuming slip boundary conditions instead of conventional stick boundary conditions,[11] which led to the theoretical data shown in Figures 3.4.4 and 3.4.5. For example, the assumption of slip boundary conditions would lead to an increase of the translational diffusion coefficient of a sphere by the factor of 1.5 as compared to the calculated values for D^T in Figures 3.4.4 and 3.4.5, whereas the rotational motion of a sphere does not displace any fluid, which implies a diverging rotational diffusion coefficient. Both results do not agree with the experimental data. Hence, the assumption of slip in contrast to stick boundary conditions is not appropriate for modeling the diffusion of the clusters. Slip boundary conditions work better for small objects of molecular dimensions which have lower orientation times than the particle clusters.[11,52]
Modeling on the basis of the shell model was used to confirm the kind of rotation which is monitored in the DDLS experiment because diffusion coefficients for the rotation around all specific axes can be calculated and compared to the experimental result. A sphere does not possess specific axes of rotation. However, particle doublets and triplets have two specific axes of rotation (Figure 3.4.5). $D_{\parallel}^R$ characterizes the rotation around the main symmetry axis, that is, the axis that connects the centers of the building blocks in the case of the doublet, whereas the main symmetry of the triplet is the C3 axis perpendicular to the plane of the constituents. $D_{\perp}^R$ is related to the rotation around the axis perpendicular to the main symmetry axis of the objects. In both cases, the rotational diffusion coefficient $D_{\perp}^R$ is measured by DDLS due to the shape anisotropy of the

particles, while $D_{\|}^{R}$ cannot be detected because of the rather small anisotropy of the spherical building blocks (Figure 3.4.5).

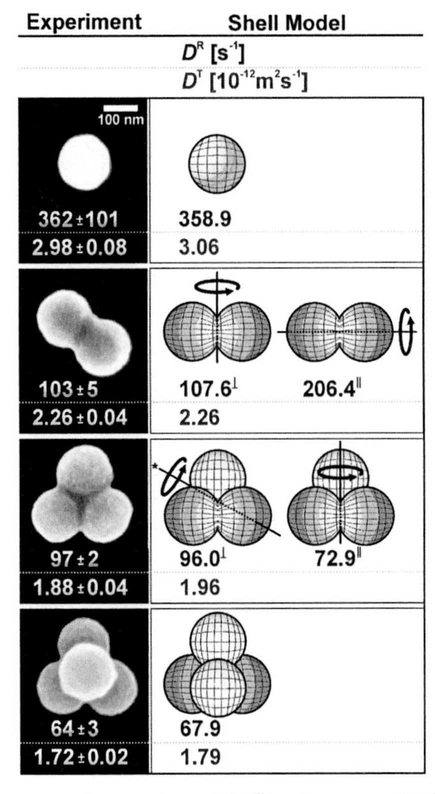

Figure 3.4.5: Comparison of the translational (D^T) and rotational ($D_{\|}^{R}$, $D_{\perp}^{R}$) diffusion coefficients as obtained by DLS and DDLS, respectively, together with the theoretical results using the shell model. For the particle doublets and triplets, the rotational diffusion coefficient $D_{\perp}^{R}$ perpendicular to the main symmetry axis is measured. In the left column, the particle clusters are oriented with their main body parallel to the plane of the figure.

As discussed in the previous section, the diffusion of a tetrahedron resembles those of a spherical object because of its low shape anisotropy. For this reason, the rotation of the tetrahedron cannot be assigned to a specific axis.

To summarize this point, the shell model gave an excellent prediction of the diffusion coefficients of the particle clusters because the true shape as derived from the FESEM micrographs could be taken into account. For this reason, the predictions obtained from this model agree well with the experimental results (Figure 3.4.5).

3.4.4 Conclusions

A common method in studying the dynamics of particles is DLS. DDLS further broadens the scope of this technique because it can be used to study both translational and rotational properties of small particles which underlie Brownian motion. Monodisperse spherical particles with diameters on the order of 100 nm can be combined to different submicron-sized clusters with well-defined shape. For this reason, and because the translational and rotational properties of submicron-sized clusters are decoupled, colloidal clusters present ideal model systems to study the diffusion of complex particles with DDLS. Unlike diffusion studies by microscopy, DDLS monitors the true diffusion properties and does not underlie wall effects. A simple description of the diffusion of clusters made up from a small number of constituents is obtained in terms of the diffusion of rotational ellipsoids. The major disadvantage is that there is no direct relation between the geometric parameters of the clusters and the rotational ellipsoids. This gap between experiment and theory can be overcome by sophisticated hydrodynamic models such as the shell model which were developed in recent years. These models allow a precise prediction of the diffusion coefficients based on the shape of the objects. Moreover, they are very useful tools to interpret experimental data sets. Hence, the present study of the diffusion of submicron-sized particle clusters contributes to the fundamental understanding of the dynamics of particles with complex shape. Hence, it is intended for general use because the dynamics of complex particles are relevant to many practical problems occurring both in nature and in industrial processes.

3.4.5 Methods

3.4.5.1. Chemicals

The chemicals used were purchased either from Sigma-Aldrich or Merck. Styrene was purified by washing with 10 wt % NaOH solution, drying over $CaCl_2$, and vacuum distillation. All other chemicals were of analytical grade and used as received.

3.4.5.2. Cluster Preparation

Amino-modified polystyrene particles were used to build the clusters. These particles were prepared by emulsion polymerization of styrene with divinylbenzene (5 mol% relative to styrene) as the crosslinker, aminoethylmethacrylate hydrochloride (AEMH, 3 mol % relative to styrene) as the comonomer, cetyltrimethylammonium bromide (CTAB) as the emulsifier, and α,α'-azodiisobutyramidine dihydrochloride (V- 50) as the initiator. The polymerization was carried out at 80 °C under a nitrogen atmosphere

and continuous stirring at 320 rpm for 6 h. Purification of the latex particles was accomplished by exhaustive ultrafiltration against water. The size and the size distribution of the spherical particles were determined by dynamic light scattering (DLS), transmission electron microscopy, and disk centrifugation. The particles have an average diameter of 154 nm and can be regarded as monodisperse because their polydispersity index given as the weight-average diameter divided by the number-average diameter is 1.004. The zeta potential of the particles bearing amino groups on their surface is 66 ± 5 mV.
The particle clusters were prepared along the lines given in ref 38. This experimental approach was based on the agglomeration of the particles which were adsorbed onto the surface of narrow-dispersed emulsion droplets. Briefly, the particles were transferred from water into toluene. Three milliliters of the 4.5 wt % suspension was added to a 0.5 wt% aqueous solution of surfactant (Pluronic F-68). A narrow-dispersed oil-in-water emulsion was obtained through powerful ultrasonication using a highshear homogenizer. The self-assembly of the particles was induced by evaporation of the toluene using a rotary evaporator.
Separation of the suspension into fractions of clusters consisting of the same number of building blocks was accomplished through density gradient centrifugation. The density gradient was prepared by using a standard gradient maker with equal volumes of a 9 and 20 wt % aqueous glycerol solution. Glycerol was used because it can be easily removed during the subsequent dialysis of the cluster fractions against water.

3.4.5.3. Methods

Field emission scanning electron microscopy (FESEM) was performed using a Zeiss LEO 1530 Gemini microscope equipped with a field emission cathode. Electrophoretic mobilities (u) of the particles were measured on a Malvern Zetasizer Nano ZS in conjunction with a Malvern MPT2 Autotitrator and converted into zeta potentials (ζ) *via* the Smoluchowski equation ($\zeta = u\eta/\epsilon_0\epsilon$, where η denotes the viscosity and $\epsilon_0\epsilon$ the permittivity of the suspension).
DLS and DDLS measurements were performed at 25 °C on a light scattering ALV/DLS/DLS-5000 compact goniometer system (Peters) equipped with a He-Ne laser (wavelength 632.8 nm), an ALV-6010/160 External Multiple Tau Digital Correlator (ALV), and a thermostat (Rotilabo). Cluster agglomerates and dust were removed through centrifugation (Beckman Coulter Allegra 64R) at 3000 rpm for 20 min. Prior to the measurement, the supernatant was filtered through 0.45 μm PET syringe filters (membra- Pure Membrex 25) into dust-free quartz glass cuvettes (Hellma). The volume fractions of the cluster suspensions were 10^{-5} to 10^{-6}. The samples were placed in a *cis*-decaline index matching bath because *cis*-decaline does not change the polarization plane of the laser light. For each sample, three runs of 180 s (DLS) or 180 to 900 s (DDLS) were performed at scattering angles of 20 to 90° (DLS) or 20 to 60° (DDLS). The scattered light passed through a Glan Thompson polarizer (B. Halle) with an extinction

ratio better than 10^{-5}. CONTIN analysis of the intensity autocorrelation functions was used to calculate the relaxation frequencies.[47]

Acknowledgment

The authors gratefully acknowledge financial support from the Deutsche Forschungsgemeinschaft (DFG) within SFB 840. M.H. thanks the Elite Network of Bavaria (ENB) within the graduate program "Macromolecular Science". A.W. is grateful to the Fonds der Chemischen Industrie (FCI), and Dr. Otto Röhm Gedächtnisstiftung.

Supporting Information Available:

Plots of the relaxation rates Γ as obtained by DLS and DDLS for single particles, doublets, and triplets. Plot showing the contribution of the depolarized signal to the total scattering intensity for the different species. This material is available free of charge *via* the Internet at http:// pubs.acs.org.

3.4.6 References and Notes

[1] Flamberg, A.; Pecora, R. Dynamic Light Scattering Study of Micelles in a High Ionic Strength Solution. *J. Phys. Chem.* **1984**, *88*, 3026–3033.

[2] Koenderink, G. H.; Zhang, H.; Aarts, D. G. A. L.; Lettinga, M. P.; Philipse, A. P.; Nägele, G. On the Validity of Stokes- Einstein-Debye Relations for Rotational Diffusion in Colloidal Suspensions. *Faraday Discuss.* **2003**, *123*, 335–354.

[3] Hoffmann, M.; Lu, Y.; Schrinner, M.; Ballauff, M.; Harnau, L. Dumbbell-Shaped Polyelectrolyte Brushes Studied by Depolarized Dynamic Light Scattering. *J. Phys. Chem. B* **2008**, *112*, 14843–14850.

[4] Shetty, A. M.; Wilkins, G. M. H.; Nanda, J.; Solomon, M. J. Multiangle Depolarized Dynamic Light Scattering of Short Functionalized Single-Walled Carbon Nanotubes. *J. Phys. Chem. C* **2009**, *113*, 7129–7133.

[5] Kanetakis, J.; Sillescu, H. Simultaneous Measurement of Rotational and Translational Diffusion by Forced Rayleigh Scattering. Colloid Spheres in Suspension. *Chem. Phys. Lett.* **1996**, *252*, 127–134.

[6] Phalakornkul, J. K.; Gast, A. P.; Pecora, R. Rotational Dynamics of Rodlike Polymers in a Rod/Sphere Mixture. *J. Chem. Phys.* **2000**, *112*, 6487–6494.

[7] Plum, M. A.; Steffen, W.; Fytas, G.; Knoll, W.; Menges, B. Probing Dynamics at Interfaces: Resonance Enhanced Dynamic Light Scattering. *Opt. Express* **2009**, *17*, 10364–10371.

[8] Heyes, D. M.; Nuevo, M. J.; Morales, J. J.; Branka, A. C. Translational and Rotational Diffusion of Model Nanocolloidal Dispersions Studied by Molecular Dynamics Simulations. *J. Phys.: Condens. Matter* **1998**, *10*, 10159–10178.

[9] Pecora, R. Quasi-Elastic Light Scattering from Macromolecules. *Annu. Rev. Biophys. Bioeng.* **1972**, *1*, 257– 276.

[10] Nägele, G. Viscoelasticity and Diffusional Properties of Colloidal Model Dispersions. *J. Phys.: Condens. Matter* **2003**, *15*, 407–414.

[11] Hu, C.-M.; Zwanzig, R. Rotational Friction Coefficients for Spheroids with the Slipping Boundary Condition. *J. Chem. Phys.* **1974**, *60*, 4354–4357.

[12] Aragon, S. R.; Pecora, R. Theory of Dynamic Light Scattering from Large Anisotropic Particles. *J. Chem. Phys.* **1977**, *66*, 2506–2516.

[13] Takagi, S.; Tanaka, H. Phase-Coherent Light Scattering Spectroscopy. II. Depolarized Dynamic Light Scattering. *J. Chem. Phys.* **2001**, *114*, 6296–6302.

[14] Degiorgio, V.; Piazza, R. Rotational Diffusion in Concentrated Colloidal Dispersions of Hard Spheres. *Phys. Rev. E* **1995**, *52*, 2707–2717.

[15] Degiorgio, V.; Piazza, R.; Corti, M.; Stavans, J. Dynamic Light Scattering Study of Concentrated Dispersions of Anisotropic Spherical Colloids. *J. Chem. Soc., Faraday Trans.* **1991**, *87*, 431–434.

[16] Koenderink, G. H.; Philipse, A. P. Rotational and Translational Self-Diffusion in Colloidal Sphere Suspensions and the Applicability of Generalized Stokes- Einstein Relations. *Langmuir* **2000**, *16*, 5631–5638.

[17] Lettinga, M. P.; Koenderink, G. H.; Kuipers, B. W. M.; Bessels, E.; Philipse, A. P. Rotational Dynamics of Colloidal Spheres Probed with Fluorescence Recovery after Photobleaching. *J. Chem. Phys.* **2004**, *120*, 4517–4529.

[18] Lettinga, M. P.; van Kats, C. M.; Philipse, A. P. Rotational Diffusion of Tracer Spheres in Packings and Dispersions of Colloidal Spheres Studied with Time-Resolved Phosphorescence Anisotropy. *Langmuir* **2000**, *16*, 6166–6172.

[19] Matsuoka, H.; Morikawa, H.; Yamaoka, H. Rotational Diffusion of Ellipsoidal Latex Particles in Dispersion as Studied by Depolarized Dynamic Light Scattering. *Colloids Surf., A* **1996**, *109*, 137–145.

[20] Quirantes, A.; Ben-Taleb, A.; Delgado, A. V. Determination of Size/Shape Parameters of Colloidal Ellipsoids by Photon Correlation Spectroscopy. *Colloids Surf., A* **1996**, *119*, 73–80.

[21] Brogioli, D.; Salerno, D.; Cassina, V.; Sacanna, S.; Philipse, A. P.; Croccolo, F.; Mantegazza, F. Characterization of Anisotropic Nano-Particles by Using Depolarized Dynamic Light Scattering in the Near Field. *Opt. Express* **2009**, *17*, 1222–1233.

[22] de Souza Lima, M. M.; Wong, J. T.; Paillet, M.; Borsali, R.; Pecora, R. Translational and Rotational Dynamics of Rodlike Cellulose Whiskers. *Langmuir* **2003**, *19*, 24–29.

[23] Koenderink, G. H.; Aarts, D. G. A. L.; Philipse, A. P. Rotational Dynamics of Colloidal Tracer Spheres in Suspensions of Charged Rigid Rods. *J. Chem. Phys.* **2003**, *119*, 4490–4499.

[24] Lehner, D.; Lindner, H.; Glatter, O. Determination of the Translational and Rotational Diffusion Coefficients of Rodlike Particles Using Depolarized Dynamic Light Scattering. *Langmuir* **2000**, *16*, 1689–1695.

[25] van Bruggen, M. P. B.; Lekkerkerker, H. N. W.; Maret, G.; Dhont, J. K. G. Long-Time Translational Self-Diffusion in Isotropic and Nematic Dispersions of Colloidal Rods. *Phys. Rev. E* **1998**, *58*, 7668–7677.

[26] Kroeger, A.; Deimede, V.; Belack, J.; Lieberwirth, I.; Fytas, G.; Wegner, G. Equilibrium Length and Shape of Rodlike Polyelectrolyte Micelles in Dilute Aqueous Solutions. *Macromolecules* **2007**, *40*, 105–115.

[27] Jabbari-Farouji, S.; Eiser, E.; Wegdam, G. H.; Bonn, D. Ageing Dynamics of Translational and Rotational Diffusion in a Colloidal Glass. *J. Phys.: Condens. Matter* **2004**, *16*, L471–L477.

[28] Carrasco, B.; Garcia de la Torre, J. Hydrodynamic Properties of Rigid Particles: Comparison of Different Modeling and Computational Procedures. *Biophys. J.* **1999**, *76*, 3044–3057.

[29] Myers, D. *Surfaces, Interfaces, and Colloids: Principles and Applications*, 2nd ed.; Wiley-VCH: New York, 1999.

[30] Anthony, S. M.; Kim, M.; Granick, S. Translation-Rotation Decoupling of Colloidal Clusters of Various Symmetries. *J. Chem. Phys.* **2008**, *129*, 244701.

[31] Kim, M.; Anthony, S. M.; Granick, S. Isomeric Colloidal Clusters with Shape-Dependent Mobility. *Soft Matter* **2009**, *5*, 81–83.

[32] Bolisetty, S.; Hoffmann, M.; Lekkala, S.; Hellweg, T.; Ballauff, M.; Harnau, L. Coupling of Rotational Motion with Shape Fluctuations of Core-Shell Microgels Having Tunable Softness. *Macromolecules* **2009**, *42*, 1264–1269.

[33] Velev, O. D.; Furusawa, K.; Nagayama, K. Assembly of Latex Particles by Using Emulsion Droplets as Templates. 2. Ball-like and Composite Aggregates. *Langmuir* **1996**, *12*, 2385–2391.

[34] Velev, O. D.; Furusawa, K.; Nagayama, K. Assembly of Latex Particles by Using Emulsion Droplets as Templates. 1. Microstructured Hollow Spheres. *Langmuir* **1996**, *12*, 2374–2384.

[35] Manoharan, V. N.; Elsesser, M. T.; Pine, D. J. Dense Packing and Symmetry in Small Clusters of Microspheres. *Science* **2003**, *301*, 483–487.

[36] Zerrouki, D.; Rotenberg, B.; Abramson, S.; Baudry, J.; Goubault, C.; Leal-Calderon, F.; Pine, D. J.; Bibette, J. Preparation of Doublet, Triangular, and Tetrahedral Colloidal Clusters by Controlled Emulsification. *Langmuir* **2006**, *22*, 57–62.

[37] Manoharan, V. N. Colloidal Spheres Confined by Liquid Droplets: Geometry, Physics, and Physical Chemistry. *Solid State Commun.* **2006**, *139*, 557–561.

[38] Wagner, C. S.; Lu, Y.; Wittemann, A. Preparation of Submicrometer-Sized Clusters from Polymer Spheres Using Ultrasonication. *Langmuir* **2008**, *24*, 12126–12128.

[39] Bantchev, G. B.; Russo, P. S.; McCarley, R. L.; Hammer, R. P. Simple Multiangle, Multicorrelator Depolarized Dynamic Light Scattering Apparatus. *Rev. Sci. Instrum.* **2006**, *77*, 043902.

[40] Eimer, W.; Dorfmueller, T. Self-Aggregation of Guanosine 5'-Monophosphate Studied by Dynamic Light Scattering Techniques. *J. Phys. Chem.* **1992**, *96*, 6790–6800.

[41] Eimer, W.; Pecora, R. Rotational and Translational Diffusion of Short Rodlike Molecules in Solution: Oligonucleotides. *J. Chem. Phys.* **1991**, *94*, 2324–2329.

[42] Eimer, W.; Williamson, J. R.; Boxer, S. G.; Pecora, R. Characterization of the Overall and Internal Dynamics of Short Oligonucleotides by Depolarized Dynamic Light Scattering and Nmr Relaxation Measurements. *Biochemistry* **1990**, *29*, 799–811.

[43] Kim, S.; Karilla, S. *Microhydrodynamics*; Butterworth-Heinemann: New York, 1991.

[44] Garcia de la Torre, J.; Del Rio Echenique, G.; Ortega, A. Improved Calculation of Rotational Diffusion and Intrinsic Viscosity of Bead Models for Macromolecules and Nanoparticles. *J. Phys. Chem. B* **2007**, *111*, 955–961.

[45] Maeda, H.; Maeda, Y. Direct Observation of Brownian Dynamics of Hard Colloidal Nanorods. *Nano Lett.* **2007**, *7*, 3329–3335.

[46] Binks, B. P.; Horozov, T. *Colloidal Particles at Liquid Interfaces*; Cambridge University Press: Cambridge, 2006.

[47] Ju, R. T. C.; Frank, C. W.; Gast, A. P. Contin Analysis of Colloidal Aggregates. *Langmuir* **1992**, *8*, 2165–2171.

[48] Schumacher, G. A.; Van de Ven, T. G. M. Brownian Motion of Rod-Shaped Colloidal Particles Surrounded by Electrical Double Layers. J. Chem. Soc., *Faraday Trans.* **1991**, *87*, 971–976.

[49] Watzlawek, M.; Nägele, G. Self-Diffusion Coefficients of Charged Particles: Prediction of Nonlinear Volume Fraction Dependence. *Phys. Rev. E* **1997**, *56*, 1258–1261.

[50] Voudouris, P.; Choi, J.; Dong, H.; Bockstaller, M. R.; Matyjaszewski, K.; Fytas, G. Effect of Shell Architecture on the Static and Dynamic Properties of Polymer-Coated Particles in Solution. *Macromolecules* **2009**, *42*, 2721–2728.

[51] Hansen, S. Translational Friction Coefficients for Cylinders of Arbitrary Axial Ratios Estimated by Monte Carlo Simulation. *J. Chem. Phys.* **2004**, *121*, 9111–9115.

[52] Fytas, G.; Wang, C. H. Studies of Siloxane Oligomers by Depolarized Rayleigh Scattering. *J. Am. Chem. Soc.* **1984**, *106*, 4392–4396.

3.4.7 Supporting Information

3D Brownian Diffusion of Submicron-sized Particle Clusters

Martin Hoffmann,[†] Claudia S. Wagner,[†] Ludger Harnau[‡,S] and Alexander Wittemann[†*]

[†] *Physikalische Chemie I, Universität Bayreuth, Universitätsstr. 30, 95440 Bayreuth, Germany*
[‡] *Max-Planck-Institut für Metallforschung, Heisenbergstr. 3, D-70569 Stuttgart, Germany,*
[S] *Institut für Theoretische und Angewandte Physik, Universität Stuttgart, Pfaffenwaldring 57, D-70569 Stuttgart, Germany*
alexander.Wittemann@uni-bayreuth.de. Telephone: +49 921 55 2776.
Fax: +49 921 55 2780

3.4.7.1. DLS and DDLS experiments

Single particles ($N = 1$)

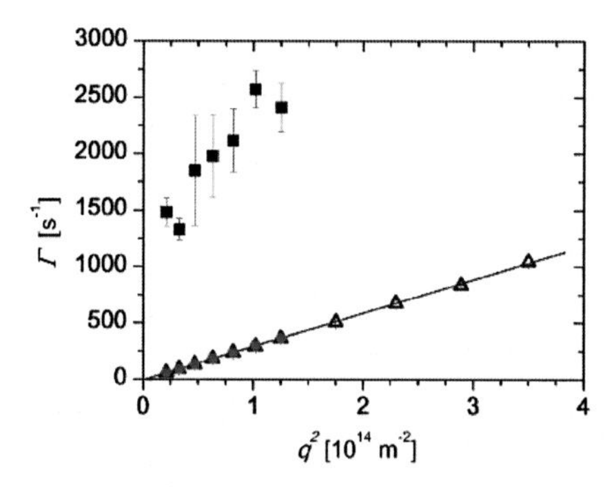

Supporting Figure 1: Relaxation rates Γ as the function of the square of the scattering vector q: The blue triangles denote the slow mode of the DLS experiment, whereas the red circles refer to the slow mode of the DDLS measurement. The fast mode of the DDLS-experiment is represented as black squares. For single particles, the contribution of the fast mode to the intensity autocorrelation function was poor. In this case, the rotational diffusion coefficient D^R was calculated for the individual values of Γ according to the following expression: $D^R = (\Gamma_{\text{fast}} - D^T q^2)/6$. D^R is then given as the average of D^R.

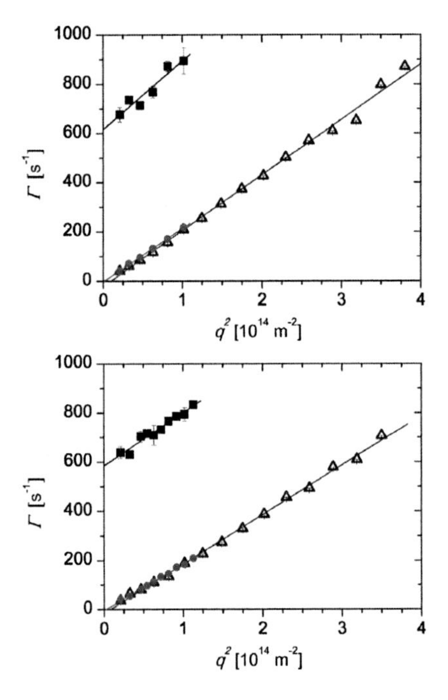

Supporting Figure 2: Relaxation rates Γ as the function of the square of the scattering vector q: The blue triangles denote the slow mode of the DLS experiment, whereas the red circles refer to the slow mode of the DDLS measurement (top: doublets; bottom: triplets). The fast mode of the DDLS experiment is represented as black squares. Linear regressions of the slow mode and the fast mode have the same slope. Hence, the translational diffusion coefficients D^T obtained by DLS and DDLS are the same within the limits of experimental errors.

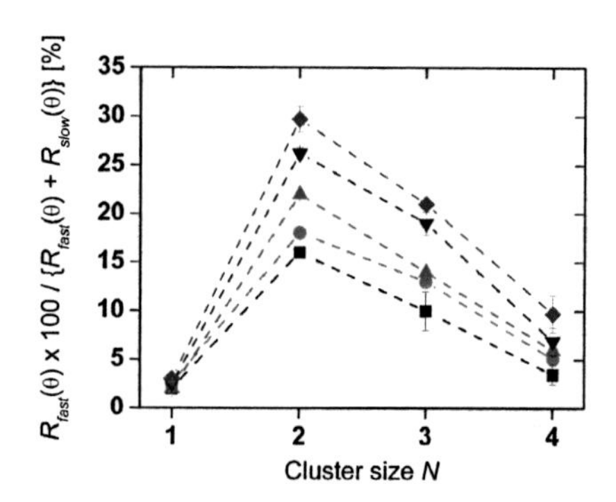

Supporting Figure 3: Contributions of the fast mode $R_{fast}(\theta)$ to the total scattered intensity $(R_{slow}(\theta) + R_{fast}(\theta))$ for scattering angles of 20° (■), 25° (●), 30° (▲), 35° (▼), and 40° (◆). The ratios $R(\theta)$ were obtained through integration of the relaxation time distributions $A(\tau)$ of the discrete modes (Figure 3.4.2C). The building blocks of the clusters ($N = 1$) and particle tetrahedrons ($N = 4$) show only low relative values of $R_{fast}(\theta)$ because of the low anisotropies of these two objects, whereas the shape anisotropies of particle doublets ($N = 2$) and triplets ($N = 3$) give rise to marked contributions of the fast mode to the scattered intensity.

Chapter 4

Summary / Zusammenfassung

Summary

This work describes the fabrication and the characterization of anisotropic colloidal particles. Spherical polyelectrolyte brushes (SPB), two stimuli-responsive dumbbell-shaped core-shell systems and submicron-sized colloidal clusters were employed to investigate the influence of particle morphology on the dynamics in solution by using (depolarized) dynamic light scattering (D)DLS.
First of all, it could be shown that the investigation of hydrodynamic and electrodynamic properties connected with a polyelectrolyte layer is fundamental for a better understanding of polyelectrolyte brushes in general. For this purpose, the influence of trivalent counterions on a SPB in an electric field was investigated. The hydrodynamic radius R_h and the electrophoretic mobility μ of the SPB were ascribed to the same hydrodynamic shear plane. From these quantities, the surface potential ζ was calculated regarding the SPB as compact colloids. All experimental quantities (R_h, μ, ζ) showed a steep decrease at the collapse transition of the brush layer when most of the monovalent counterions were exchanged by trivalent ones. Variational free energy calculations for an isolated SPB corroborated the experimental findings. The results strongly indicate that electroosmotic flows can be widely suppressed in a polyelectrolyte brush due to the strong correlation of multivalent counterions to the polyelectrolyte chains.
To obtain non-spherical polyelectrolyte brushes, a preparation pathway for stable dumbbell-shaped polyelectrolyte brushes (DPB) was established. The core particles were made of poly(methyl methacrylate) and poly(styrene). Their surface could be modified by a dense layer of grafted poly(styrene sulfonate) chains in a photoemulsion polymerization. The dumbbell-morphology was proved by field emission scanning electron microscopy and cryogenic-transmission electron microscopy. A detailed DDLS analysis of the DPB at different ionic strength revealed that the overall size and dynamics of the DPB depend markedly on the brush layer thickness. For both the DPB and a spherical reference system (SPB), DDLS revealed an unexpected relaxation process in addition to

the translational and rotational diffusion coefficient. This relaxation mode could be assigned to a collective motion of the brush layer. The relaxation mode was interpreted as the dynamic counterpart of the static scattering intensities which were measured earlier in the case of SPB due to spatial fluctuations of the grafted chains.
In a further study, dumbbell-shaped core-shell microgels (DMP) were synthesized to obtain a model system with both anisotropic and thermoresponsive properties. The DMP present a new class of colloidal molecules with a temperature dependent shape and aspect ratio. For their fabrication, a seeded particle emulsion polymerization approach was modified, which had been developed for the simpler spherical analogue. FESEM and scanning force microscopy verified the charge induced 2-dimensional self-assembly of the DMP on a Si and a glass substrate. DDLS confirmed the acceleration of the translational and the rotational diffusion upon increasing the temperature. The application of the shell model gave the layer thickness, which was in very good agreement with microscopic evidence (cryo-TEM). It could be shown that the layer thickness was the same for the DMP and a spherical reference system within the experimental error. This is important to apply the synthetic approach independently of the core particle shape.
Based on the previous findings, an advanced DDLS study with monodisperse colloidal clusters gave detailed insight how particle geometry is related to the experimental diffusion coefficients for translational (D^T) and rotational (D^R) motion. It was found that the decay of D^T widely followed from the increase in the mean radius with the number of consituent spheres ($N = 1-4$). Applying the shell model clearly revealed that DDLS only detects the rotational relaxation around the minor particle axis for $N = 2$ and around the axis in the plane of the triplet for $N = 3$. Furthermore, it was shown that the rotation for $N = 4$ cannot be assigned to a specific axis since the diffusion of the tetrahedron resembles that of a sphere due to its low shape anisotropy. The DDLS signal for $N = 1$ was attributed to the small optical anisotropy of the building blocks.

Zusammenfassung

In dieser Arbeit wird die Herstellung und Charakterisierung von anisotropen Kolloiden beschrieben. Der Einfluss der Teilchenmorphologie auf die Dynamik in Lösung wurde durch den Einsatz von (depolarisierter) dynamischer Lichtstreuung (D)DLS an sphärischen Polyelektrolytbürsten untersucht, an zwei hantelförmigen Kern-Schale Systemen, die auf externe Stimuli reagieren, sowie an kolloidalen Partikelclustern von der Größe einiger hundert Nanometer.
Zunächst wurde gezeigt, dass die Untersuchung der hydrodynamischen und elektrodynamischen Eigenschaften einer Polyelektrolytschicht zum vertieften Verständnis von Polyelektrolytbürsten beiträgt. Hierzu wurde der Einfluss dreiwertiger Gegenionen auf sphärische Polyelektrolytbürsten in einem elektrischen Feld untersucht. Für den hydrodynamischen Radius R_h und die elektrophoretische Mobilität μ der SPB Partikel wurde die

gleiche hydrodynamische Scherebene vorausgesetzt. Das Oberflächenpotential ζ wurde unter der Annahme berechnet, dass sich die SPB wie kompakte Kolloide verhalten. Alle experimentellen Größen (R_h, μ, ζ) nahmen beim Kollaps der Polyelektrolytbürste stark ab. Dies ist der Fall, wenn der Austausch von ein- gegen dreiwertige Gegenionen in der Bürste fast vollständig ist. Die Resultate wurden durch Variationsrechnung für die freie Energie eines isolierten SPB bestätigt. Wie die Daten zeigten, können auf Grund der starken Korrelation mehrwertiger Gegenionen mit Polyelektrolytketten elektroosmotische Ströme in einer Polyelektrolytbürste stark erniedrigt werden.
Um nicht-sphärische Polyelektrolytbürsten zu erhalten, wurde eine Syntheseroute für stabile hantelförmige Polyelektrolytbürsten (DPB) ausgearbeitet. Die Kernpartikel bestanden aus Polymethylmethacrylat und Polystyrol. Ihre Oberfläche wurde mittels Photoemulsionspolymerisation mit einer dichten Schicht von Ketten aus Polystyrolsulfonat modifiziert. Die hantelförmige Gestalt wurde mittels Feldemissionsrasterelektronenmikroskopie (FESEM) und Tieftemperatur-Transmissionselektronenmikroskopie (cryo-TEM) nachgewiesen. Wie eine detaillierte Untersuchung der DPB Partikel mittels DDLS bei unterschiedlichen Ionenstärken zeigte, hängen Größe und Dynamik der DPB maßgeblich von der Dicke der Bürstenschicht ab. Für die DPB Partikel sowie ein sphärisches Referenzsystem (SPB) wurde mittels DDLS zusätzlich zu Translations- und Rotationsdiffusion ein unerwarteter Relaxationsprozess gefunden. Diese Relaxationsmode konnte mit einer kollektiven Bewegung der Polyelektrolytbürste erklärt werden. Die Mode lässt sich als dynamisches Analogon zur statischen Streuintensität verstehen, wie sie bereits an SPB aufgrund von räumlichen Fluktuationen der angebundenen Ketten gemessen wurde.
In einer anschließenden Untersuchung wurden hantelförmige Mikrogele mit Kern-Schale Architektur (DMP) hergestellt, um ein Modellsystem mit anisotropen und temperatursensitiven Eigenschaften zu erhalten. Die DMP bilden eine neue Klasse von kolloidalen Molekülen, die auf Temperaturveränderungen mit Anpassung von Form und Aspektverhältnis reagieren. Hierfür wurde ein Verfahren zur Emulsionspolymerisation mit Saatpartikeln weiter entwickelt, welches ursprünglich für einfachere sphärische Kern-Schale Partikel konzipiert worden war. Mit FESEM und Rasterkraftmikroskopie wurde die ladungsinduzierte zweidimensionale Selbstanordnung der DMP auf einem Silizium- und einem Glasssubstrat untersucht. Mittels DDLS wurde die Beschleunigung der Translations- und Rotationsbewegung bei Temperaturerhöhung bestätigt. Die Anwendung des Schalenmodelles erlaubte die Berechnung der Schalendicke. Diese stimmte mit Befunden aus bildgebenden Verfahren (cryo-TEM) sehr gut überein. Es konnte gezeigt werden, dass die Schalendicken für die DMP und ein sphärisches Referenzsystem im Rahmen des Fehlers identisch sind. Dies ist von Bedeutung für die Anwendung des präparativen Ansatzes auf Kerne unterschiedlicher Form.
Weitergehende Untersuchungen an monodispersen kolloidalen Clustern lieferten Informationen über den Zusammenhang von Partikelform und den entsprechenden experi-

mentell zugänglichen Diffusionskoeffizienten für Translation (D^T) und Rotation (D^R). Wie sich zeigte, wird die Abnahme von D^T weitgehend vom Anstieg des mittleren Teilchenradius mit der Zahl der am Clusteraufbau beteiligten sphärischen Bausteine bestimmt ($N = 1 - 4$). Die Anwendung des Schalenmodells zeigte eindeutig, dass im DDLS Experiment lediglich die Rotationsrelaxation um die kürzere Achse im Fall von $N = 2$ gemessen wird, und um die Achse in der Ebene des Tripletts für $N = 3$. Weiterhin konnte gezeigt werden, dass im Fall $N = 4$ die Rotation keiner bestimmten Achse zugeordnet werden kann, da die Diffusion eines Tetraeders auf Grund der geringen Formanisotropie der einer Kugel ähnelt. Das DDLS Signal der Einzelkugel ($N = 1$) wurde mit der niedrigen optischen Anisotropie der Bausteine erklärt.

Chapter 5

List of Publications

[1] Mei, Y.; Lauterbach, K.; **Hoffmann, M.**; Borisov, O.; Ballauff, M.; Jusufi, A.: Collapse of Spherical Polyelectrolyte Brushes in the Presence of Multivalent Counterions, *Phys. Rev. Lett.* **2006**, *97*, 158301.

[2] Walther, A.; **Hoffmann, M.**; Müller, A. H. E.: Emulsion Polymerization Using Janus Particles as Stabilizers, *Angew. Chem. Int. Ed.* **2008**, *47*, 711-714.

[3] Mei, Y.; **Hoffmann, M.**; Ballauff, M.; Jusufi, A.: Spherical polyelectrolyte brushes in the presence of multivalent counterions: The effect of fluctuations and correlations as determined by molecular dynamics simulations, *Phys. Rev. E* **2008**, *77*, 031805.

[4] **Hoffmann, M.**; Lu, Y.; Schrinner, M.; Ballauff, M.; Harnau, L.: Dumbbell-Shaped Polyelectrolyte Brushes Studied by Depolarized Dynamic Light Scattering. *J. Phys. Chem. B.* **2008**, *112*, 14843-14850.

[5] Lu, Y.; **Hoffmann, M.**; Yelamanchili, R. S.; Terrenoire, A.; Schrinner, M.; Drechsler, M.; Möller, M. W.; Breu, J; Ballauff, M.: Well-defined crystalline TiO2-Nanoparticles Generated and Immobilized on a Colloidal Nanoreactor. *Macromol. Chem. Phys.* **2009**, *210*, 377-386.

[6] Bolisetty, S.; **Hoffmann, M.**; Hellweg, T.; Harnau, L.; Ballauff, M.: Coupling of Rotational Motion with Shape Fluctuations of Core-Shell Microgels Having Tunable Softness. *Macromolecules* **2009**, *42*, 1264-1269.

[7] **Hoffmann, M.**; Jusufi, A.; Schneider, C.; Ballauff, M.: Surface potential of spherical polyelectrolyte brushes in the presence of trivalent counterions. *J. Coll. Interf. Sci.* **2009**, *338*, 566-572.

[8] **Hoffmann, M.**; Wagner, C. S.; Harnau, L.; Wittemann, A.: 3D Brownian Diffusion of Submicron-Sized Particle Clusters, *ACS NANO*, **2009**, *3*, 3326-3334.

[9] **Hoffmann, M.**; Siebenbürger, M.; Harnau, L.; Hund, M.; Hanske, C.; Lu, Y.; Wagner, S.; Drechsler, M.; Ballauff, M.: Thermoresponsive Colloidal Molecules, *Soft Matter*, **2010**, *6*, 1125-1128.

Chapter 6

Conference Presentations and Workshops

(1) Martin Hoffmann, Yan Lu, Marc Schrinner, Matthias Ballauff and Ludger Harnau, *"Synthesis and Characterization of Dumbbell-Shaped Polyelectrolyte Brush Particles"*, poster presentation at the conference Makromolekulares Kolloquium 2008, February 28 - March 1, 2008, Freiburg (GER)

(2) Martin Hoffmann, Yan Lu, Marc Schrinner, Matthias Ballauff and Ludger Harnau, *"Synthesis and Characterization of Dumbbell-Shaped Polyelectrolyte Brush Particles"*, poster presentation at ECIS conference, August 31 - September 5, 2008, Krakow (POL)

(3) *School of Surface Analytical Techniques*, workshop at the Max-Planck Institute of Colloids and Interfaces, March 11 - 14, 2008, Golm (GER)

Appendix A

Index

F

G

H

L

M

O

P

R

S

V

Appendix B

Abbreviations

Symbol	Meaning
$< N >$	Average number of particles in the scattering volume (m^{-3})
α	Aspect ratio
α_{aniso}	Optical anisotropy ($C^2 \cdot m^2 \cdot J^{-1}$)
α_{iso}	Isotropic part of the polarizability tensor ($C^2 \cdot m^2 \cdot J^{-1}$)
χ_{mp}	Monomer-polymer interaction parameter
$\Delta\overline{G}$	Gibbs free energy difference ($J \cdot mol^{-1}$)
η	Viscosity ($kg \cdot m^{-1} \cdot s^{-1}$)
Γ	Relaxation rate (s^{-1})
γ	Interfacial tension ($J \cdot m^{-2}$)
λ	Wavelength of the laser (632.8 nm)
$\mathcal{M}_2$	Second moment of the mass distribution (m^2)
μ	Electrophoretic mobility ($m^2 \cdot V^{-1} \cdot s^{-1}$)
$\overline{P_w}/\overline{P_n}$	Polydispersity of polymer chains
$\parallel$	Parallel
$\perp$	Perpendicular
ϕ	Volume fraction (%)
Ψ	Surface potential (V)
σ	Grafting density, number of chains per unit area (nm^{-2})
θ	Scattering angle (°)
v_p	Volume fraction of polymer in a monomer-swollen, crosslinked particle (%)
$\vec{r}_0$	Center of mass of a colloidal cluster
$\vec{r}_i$	Center of a building block in a cluster
ζ	Electrokinetic potential, zeta-Potential (V)
a	Longer semiaxis of a prolate ellipsoid (m)
a_{CL}	Radius of a crosslinked polymer particle swollen with monomer
b	Shorter semiaxis of a prolate ellipsoid (m)
d	Diameter of a cylinder (m)

D^R .... Rotational diffusion coefficient (s^{-1})
D^T .... Translational diffusion coefficient ($m^2\ s^{-1}$)
e .... elementary unit ($1.660\,22 \cdot 10^{-19}$C)
$G(\Gamma)$ .... Relaxation rate distribution function
$g^{(1)}(q,t)$ .... Field-field autocorrelation function
$g^{(2)}(q,t)$ .... Intensity-intensity autocorrelation function
L .... Length of a cylinder (m)
l .... Center-to-center distance in a double sphere (m)
L_c .... Contour length (m)
L_h .... Thickness of the microgel layer (m)
N .... Number of building blocks in a colloidal cluster
n .... Refractive index
N_c .... Effective number density of chains in a polymer-network ($mol \cdot m^{-3}$)
q .... $= 4\pi \sin(\theta/2) n/\lambda$ absolute value of the scattering vector (m^{-1})
Q^* .... Effective particle charge (e)
R .... Universal gas constant ($8.3145\ J \cdot mol^{-1} \cdot K^{-1}$)
R_c .... Core radius of a spherical polyelectrolyte brush (m)
$S(q,t)$ .... Dynamic form factor
t .... time (s)
V_m .... Molar volume of a monomer ($m^3 \cdot mol^{-1}$)
vH .... vertical horizontal
vV .... vertical vertical
ASAXS .... Anomalous small-angel X-ray scattering
BIS .... N,N'-methylenebisacrylamide
cryo-TEM ...... Cryogenic-transmission electron microscopy
DDLS .... Depolarized dynamic light scattering
DMP .... Dumbbell-shaped thermoresponsive microgel
DPB .... Dumbbell-shaped polyelectrolyte brush
fast .... Denotes the relaxation rate as measured by DDLS
FESEM .... Field emission scanning electron microscopy
HMEM .... 2-[p-(2-hydroxy-2-methylpropiophenone)] - ethylenglycol methacrylat
KPS .... Potassimum peroxodisulfate
LCST .... Lower critical solution temperature (°C)
NaSS .... Sodium styrene sulfonate
NDPyCl .... N-dodecylpyridinium chloride
PMMA .... poly(methyl methacrylate)
PNIPA .... poly(N-isopropylacrylamide)
PS .... poly(styrene)
PSS .... poly(styrene sulfonate)
SAXS .... Small-angle X-ray scattering

SDBS .... Sodium dodecylstyrene sulfonate
SFM .... Scanning force microscopy
slow .... Denotes the slow relaxation rate measured by DLS and DDLS
SPB .... Spherical polyelectrolyte brush
TEM .... Transmission electron microscopy
theo .... theoretical

Appendix C

Copyright and Permission Notices

Figures and Schemes

- Figure 1.1.2 on p 4 was reprinted with permission from:
  Yu Mei, Karlheinz Lauterbach, Martin Hoffmann, Oleg V. Borisov, Matthias Ballauff and Arben Jusufi, Phys. Rev. Lett. 97, 158301 (2006). Copyright 2006 by the American Physical Society.

- Figure 1.2.1a on p 6 was reprinted with permission from:
  Jin-Woong Kim, Ryan J. Larsen and David A. Weitz, J. Am. Chem. Soc., 2006, 128 (44), pp 14374-14377. Copyright 2006 American Chemical Society.

- Figure 1.2.1b on p 6 was reprinted from:
  Journal of Colloid and Interface Science 328 (1), pp 98-102, Wei Lua, Min Chena and Limin Wu, One-step synthesis of organic-inorganic hybrid asymmetric dimer particles via miniemulsion polymerization and functionalization with silver, Copyright 2008, with permission from Elsevier.

- Figure 1.2.1c on p 6 was reprinted with permission from:
  Patrick M. Johnson, Carlos M. van Kats and Alfons van Blaaderen, Langmuir, 2005, 21 (24), pp 11510-11517. Copyright 2005 American Chemical Society.

- Figure 1.2.1d on p 6 was reprinted with permission from:
  Stéphane Reculusa, Céline Poncet-Legrand, Adeline Perro, Etienne Duguet, Elodie Bourgeat-Lami, Christophe Mingotaud and Serge Ravaine, Chem. Mater., 2005, 17 (13), pp 3338-3344. Copyright 2005 American Chemical Society.

- Figure 1.2.2a on p 7 was reprinted with permission from:
  Jin-Woong Kim, Ryan J. Larsen, and David A. Weitz, J. Am. Chem. Soc., 2006, 128 (44), pp 14374-14377. Copyright 2006 American Chemical Society.

- Figure 2.2.1 on p 28 was reprinted with permission from:
Martin Hoffmann, Yan Lu, Marc Schrinner and Matthias Ballauff, J. Phys. Chem. B, 2008, 112 (47), pp 14843-14850. Copyright 2008 American Chemical Society.

- Figure 2.3.1 on p 29 was reproduced by permission of The Royal Chemical Society:
Martin Hoffmann, Miriam Siebenbürger, Ludger Harnau, Markus Hund, Christoph Hanske, Yan Lu, Claudia S. Wagner, Markus Drechsler and Matthias Ballauff, Soft Matter, 2010, 6, 1125-1128.

- Figure 2.3.2 on p 30 was reproduced by permission of The Royal Chemical Society:
Martin Hoffmann, Miriam Siebenbürger, Ludger Harnau, Markus Hund, Christoph Hanske, Yan Lu, Claudia S. Wagner, Markus Drechsler and Matthias Ballauff, Soft Matter, 2010, 6, 1125-1128.

- Figure 2.4.1 on p 31 was reprinted with permission from:
Martin Hoffmann, Claudia S. Wagner, Ludger Harnau and Alexander Wittemann, ACS Nano, 2009, 3 (10), pp 3326-3334. Copyright 2009 American Chemical Society.

- Figure 2.4.2 on p 32 was reprinted with permission from:
Martin Hoffmann, Claudia S. Wagner, Ludger Harnau and Alexander Wittemann, ACS Nano, 2009, 3 (10), pp 3326-3334. Copyright 2009 American Chemical Society.

Full Papers

- The full paper on pp 37 - 58 in Chapter 3.1 was reprinted from:
Journal of Colloid and Interface Science 338 (2), pp 566-572, M. Hoffmann, A. Jusufi, C. Schneider and M. Ballauff, Surface potential of spherical polyelectrolyte brushes in the presence of trivalent counterions , Copyright 2009, with permission from Elsevier.

- The full paper on pp 59 - 80 in Chapter 3.2 was reprinted with permission from:
Martin Hoffmann, Yan Lu, Marc Schrinner and Matthias Ballauff, J. Phys. Chem. B, 2008, 112 (47), pp 14843-14850. Copyright 2008 American Chemical Society.

- The full paper on pp 81 - 95 in Chapter 3.3 was reproduced by permission of The Royal Chemical Society:
Martin Hoffmann, Miriam Siebenbürger, Ludger Harnau, Markus Hund, Christoph Hanske, Yan Lu, Claudia S. Wagner, Markus Drechsler and Matthias Ballauff, Soft Matter, 2010, 6, 1125-1128. The article can be found on the internet under: http://dx.doi.org/10.1039/C000434K.

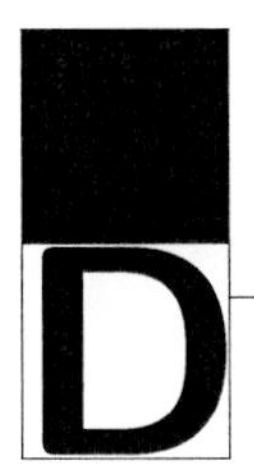

Appendix D

Danksagung

Die Zukunft soll man nicht voraussehen wollen, sondern möglich machen.
(Antoine de Saint-Exupéry)

DIESE ARBEIT wurde am Lehrstuhl für Physikalische Chemie I, Universität Bayreuth (Deutschland), in der Zeit von Oktober 2007 bis März 2010 angefertigt. An dieser Stelle sei allen herzlich gedankt, die auf großartige Weise zum Gelingen dieser Arbeit beigetragen haben:

- Meinem Doktorvater und Mentor Herrn Prof. Dr. Matthias Ballauff. An die Zeit in seiner Bayreuther Gruppe werde ich gerne zurückdenken als eine spannende und lehrreiche. Sein großes Vertrauen in mich, die Doktorarbeit in vielerlei Hinsicht selbst gestalten zu können, zusammen mit seinem fortwährenden Interesse am Fortgang der Forschung haben mich besonders beeindruckt.
- Priv.-Doz. Dr. Ludger Harnau für die hochproduktive Zusammenarbeit bei Fragestellungen zur hydrodynamischen Modellierung.
- Dr. Alexander Wittemann und Dr. Yan Lu für ihre profunden Hinweise beim Beschreiten neuer Synthesewege.
- Meinen einzigartigen Eltern, meiner Schwester und Familie für ihre Liebe und Unterstützung während meiner langen Ausbildungszeit. Dass ich immer selbst entscheiden konnte, und euer Vertrauen mich gestärkt hat.
- Thomas, Karl, Christian, Yvonne, Dominik, Tobias - für euere ehrliche Freundschaft.
- Dr. Sabine Rosenfeldt für die praktischen Ratschläge und ihre Freundschaft.

- Elisabeth Düngfelder für die schnelle und kompetente Hilfe beim bürokratischen Alltag (Chemikalienbestellungen, Dienstreiseanträge, ...).
- Meinen Kollegen am Lehrstuhl Christian Schneider, Miriam Siebenbürger, Katja Henzler und Björn Haupt, Frank Polzer, Sreenath Bolisetty und alle die ich noch nicht genannt habe für die prima Arbeitsatmosphäre.
- Christa Bächer, Christine Thunig und Karlheinz Lauterbach für ihre stete Hilfsbereitschaft und Unterstützung.
- Meinen (Vertiefungs-)praktikanten und HiWis für ihre Mitarbeit: Wolfgang Naehr, Christoph Otto Hollfelder, Tobias Kemnitzer und Franziska Heger.
- Allen Kollegen an den Lehrstühlen für Anorganische, Organische, Physikalische und Makromolekulare Chemie für die unkomplizierte Zusammenarbeit, die durch kurze Wege viele Ergebnisse beschleunigt hat.
- Dem Freistaat Bayern für die Gewährung eines Graduiertenstipendiums nach dem BayEFG und allen beteiligten Professoren und Assistenten für die hervorragende Betreuung und Möglichkeiten zur Weiterbildung im Rahmen des Studienprogrammes "Macromolecular Science".

Appendix E

Erklärung

Hiermit erkläre ich, dass ich die vorliegende Arbeit selbständig verfasst und keine anderen als die von mir angegebenen Quellen und Hilfsmittel benutzt habe.

Ferner erkläre ich, dass ich weder an der Universität Bayreuth noch an einer anderen Hochschule versucht habe, eine Dissertation einzureichen oder mich einer Promotionsprüfung zu unterziehen.

Bayreuth, den 15. März 2010

(Martin Hoffmann)